Lecture Notes in Mathematics

Edited by A. Dold and B. Eckmann

400

A Crash Course on Kleinian Groups

Lectures given at a special session at the January 1974 meeting of the American Mathematical Society at San Francisco

Edited by Lipman Bers and Irwin Kra

Springer-Verlag
Berlin · Heidelberg · New York 1974

Lipman Bers
Columbia University, Morningside Heights,
New York, NY/USA

Irwin Kra
SUNY at Stony Brook, Stony Brook. New York, NY/USA

Library of Congress Cataloging in Publication Data

American Mathematical Society.
A crash course on Kleinian groups, San Francisco, 1974.

(Lecture notes in mathematics, 400)
1. Kleinian groups. I. Bers, Lipman, ed. II. Kra, Irwin, ed. III. Title. IV. Series: Lecture notes in mathematics (Berlin, 400)
QA3.L28 no. 400 [QA331] 510'.8s [512'.55] 74-13853

AMS Subject Classifications (1970): Primary: 30-02, 32G15
Secondary: 30A46, 30A58, 30A60

ISBN 3-540-06840-6 Springer-Verlag Berlin · Heidelberg · New York
ISBN 0-387-06840-6 Springer-Verlag New York · Heidelberg · Berlin

Offsetdruck: Julius Beltz, Hemsbach/Bergstr.

To Lars V. Ahlfors

PREFACE

It has recently become customary to have "special sessions" at meetings of the AMS, consisting of short invited lectures, and intended for groups of specialists. At the Annual Winter Meeting at San Francisco, we tried to have a special session addressed to non-specialists. The lecturers were asked to prepare in advance texts of their talks, and these were distributed at the meeting. These texts, slightly revised, are collected in the present fascicule. (We also included an abstract of a forthcoming paper by H. Masur.)

The present "crash course" does not intend to do more than to give a reader an introductory survey of some topics which became important in the modern theory of Kleinian groups. The references to literature, though by no means complete, should enable anyone interested in more detailed information to obtain same.

Lars Ahlfors, who played a decisive part in the recent revival of Kleinian groups, could not be present at San Francisco. It is fitting to dedicate this modest effort to him.

L.B.

I.K.

CONTENTS

1. WHAT IS A KLEINIAN GROUP?

Lipman Bers
Columbia University

This is the first of a series of lectures which, it is hoped, will give a picture, albeit an incomplete one, of the present state of the theory of Kleinian groups.

Such groups can be studied either for their own sake (they are, by the way, the only extensively studied class of discrete subgroups of a Lie group with quotients of infinite volume) or as a tool for representing Riemann surfaces. Of course, the two points of view cannot be neatly separated.

The theory of Kleinian groups was founded by Schottky, Poincaré and Klein in the 19th century. For many years it was dormant, except, of course, for the important special case of Fuchsian groups. The burst of activity during the last decade is based, directly or indirectly, on the use of quasiconformal mappings as a working tool in complex function theory.

Since Ahlfors' seminal 1965 paper [3] finitely generated groups are at the center of attention, and our lectures will deal exclusively with such groups. Infinitely generated groups are also of interest and present new phenomena (Abikoff [1]), but applications to compact Riemann surfaces and algebraic curves (and to higher dimensional algebraic varieties, cf. Griffiths [10]) involve primarily finitely generated groups. Discontinuous groups of Möbius transformations in R^n, $n > 3$, are not discussed. Their theory is as yet in infancy and seems to have no function theoretical interest.

Since this is the first lecture it contains mostly definitions and examples.

A group of topological self-mappings of a space is called (<u>properly</u>) <u>discontinuous</u> if no compact set meets infinitely many of its translates. A <u>Kleinian group</u> G is a subgroup of the (complex) <u>Möbius group</u> Möb , which acts discontinuously on some open subset of the Riemann sphere $\hat{\mathbb{C}} = \mathbb{C} \cup \{\infty\}$. (General references: [7], [9], [11].)

Recall that $\text{Möb} = SL(2,\mathbb{C})/\{\pm I\}$ is the group of all complex unimodular 2 by 2 matrices, determined up to multiplication by (-1). The element $\pm\begin{pmatrix} a & b \\ c & d \end{pmatrix}$ acts on the complex projective line $= \hat{\mathbb{C}}$ by the rule

(1) $$z \longmapsto \frac{az+b}{cz+d} \quad ;$$

thus Möb can be identified with the group of all holomorphic isomorphisms of $\hat{C}$.

The real unimodular matrices $\pm\begin{pmatrix} a & b \\ c & d \end{pmatrix}$ form the <u>real</u>

Möbius group $\text{Möb}_{\mathbb{R}} \subset \text{Möb}$. Every element of $\text{Möb}_{\mathbb{R}}$ maps the upper half-plane $U = \{z = x + iy \in \mathbb{C},\ y > 0\}$ onto itself. Recall that the Poincaré line element $ds^2 = |dz|^2/y^2$ makes U into a model of the non-Euclidean (Bolyai-Lobatchevski) plane. Thus $\text{Möb}_{\mathbb{R}}$ may be viewed as the group of all conformal self-mappings of U and also as the group of all non-Euclidean motions in the plane.

The complex Möbius group admits a similar interpretation. The upper half-space $\mathbb{R}^3_+ = \{(z,t) |\ z \in \mathbb{C},\ t \in \mathbb{R},\ t > 0\}$ with the Poincaré metric $ds^2 = (|dz|^2 + dt^2)/t^2$ is a model of the non-Euclidean space. We use quaternions and identify a complex number $x + iy$ with the quaternion $x + iy + j0 + k0$, a point $(z,t) \in \mathbb{R}_3^+$ with the quaternion $z + jt = x + iy + jt + k0$. A complex unimodular matrix $\pm\binom{ab}{cd}$ acts on $\mathbb{R}^3_+$ by the rule

$$(2) \qquad (z + jt) \longmapsto (z' + jt') = [a(z + jt)+ b][c(z + jt)+ d]^{-1} .$$

Now Möb becomes the group of all non-Euclidean motions in space.

A discrete subgroup $G \subset \text{Möb}$ always acts discontinuously on $\mathbb{R}^3_+$ and the quotient $\mathbb{R}^3_+/G$ is always a 3-manifold (see Marden's lecture). On the other hand, G may or may not act discontinuously on an open subset of $\mathbb{C}$. If it does it is called Kleinian, as noted above, and the largest open set on which G acts discontinuously is denoted by $\Omega = \Omega(G)$

and is called the region of discontinuity. The complement $\Lambda = \Lambda(G) = \hat{\mathbb{C}}\setminus\Omega(G)$ is called the limit set of G. It is also the set of accumulation points of orbits of points, and the closure of the set of fixed points of loxodromic (including hyperbolic) and parabolic elements of G.

The limit set Λ may be finite (in this case it consists of 0, 1 or 2 points and G is called elementary) or infinite. In the latter case Λ is a perfect nowhere set of positive logarithmic capacity.

The quotient Ω/G is a 2-manifold and is given a complex structure by requiring the projection $\Omega \to \Omega/G$ to be holomorphic. Thus the quotient is a disjoint union of Riemann surfaces, $\Omega/G = S_1 + S_2 + \cdots$. (One says: G represents $S_1, S_2, \cdots$.) The projection $\Omega \to \Omega/G$ is locally one-to-one, except at points $z \in \Omega$ with non-trivial stabilizer G_z. In this case G_z is a cyclic group of some finite order ν, the projection is ν-to-1 near z, and the image of z under the projection is called a ramification point of order ν. Thus a component of Ω/G is not just a Riemann surface, but one equipped with a discrete set of ramification points, i.e., points with integers $\nu > 1$ attached to them.

If Δ is a component of Ω, the stabilizer G_Δ of Δ is again a Kleinian group, and Δ/G_Δ is a Riemann surface (with ramification points). Two components, Δ_1 and Δ_2, are called <u>conjugate</u> if $\Delta_2 = g(\Delta_1)$ for some $g \in G$. If $\Delta_1, \Delta_2, \cdots$ is a complete list of non-conjugate components of G (i.e., of $\Omega(G)$), then

(3) $$\Omega/G = \Delta_1/G_{\Delta_1} + \Delta_2/G_{\Delta_2} + \cdots .$$

The group G is said to be <u>of finite type over a component</u> Δ if Δ/G_Δ is obtained from a compact surface of genus p by removing finitely many, say $n_\infty \geq 0$, points (punctures) and if there are finitely many, say $n_0 \geq 0$, ramification points on Δ/G_Δ, of orders $\nu_1 \leq \nu_2 \leq \cdots \leq \nu_{n_0}$. Set $n = n_0 + n_\infty$. The pair (p,n) is called the <u>type</u> of Δ/G_Δ, and the sequence

(4) $$(p, n; \nu_1, \cdots, \nu_{n_0}, \underbrace{\infty, \cdots, \infty}_{n_\infty \text{ times}})$$

is called the <u>signature</u>. (One may write (p) instead of $(p,0)$.)

The group G is called <u>of finite type</u> if it has only finitely many non-conjugate components and is of finite type over all of them.

Elementary Kleinian groups are easily enumerated. The most important are the <u>cyclic</u> and <u>elliptic</u> groups. For the cyclic group $G = \{g^k, k = 0, \pm 1, \cdots\}$, Λ is empty if $g = \text{id}$ or if g is elliptic of order k, Λ consists of 1 point if g is parabolic, of 2 points if G is loxodromic. The signatures of Ω/G are (0), $(0, 2; k, k)$, $(0, 2; \infty, \infty)$ and (1), respectively. An elliptic group $G = \{z \mapsto z + nw + mw', (n,m) \in \mathbf{Z}^2, \operatorname{Im} \bar{w}w > 0\}$ has one limit point ∞; Ω/G has signature (1).

If G is non-elementary, as we assume from now on, every component Δ of Ω carries a Poincaré metric $ds^2 = \lambda(z)^2 |dz|^2$, the unique complete conformal Riemannian metric of curvature (-1). Since G respects the metric, the latter can be transplanted to Ω/G. The group G is of finite type over a component Δ if and only if

(5) $$\text{Area}\ (\Delta/G_\Delta) = \iint_{\Delta/G_\Delta} \lambda(z)^2\, dxdy < \infty ;$$

if so

(6) $$\text{Area}\ (\Delta/G_\Delta) = 2\pi\{2p - 2 + \sum_1^n (1 - \frac{1}{\nu_j})\}$$

where $(p, n; \nu_1, \ldots, \nu_n)$ is the signature of Δ/G_Δ and $1/\infty = 0$.

A basic result in the recent theory of Kleinian groups is the finiteness theorem (Ahlfors [3]) which states that a finitely generated Kleinian group is of finite type. (The converse need not be true.) A quantitative refinement are the area theorems [5]:

(7) $\text{Area } (\Omega/G) \leq 4\pi(N-1)$

if G has N generators and is not elementary,

(8) $\text{Area } (\Omega/G) \leq 2 \text{ Area } (\Delta_0/G)$ if $G_{\Delta_0} = G_\Delta$

for a non-elementary finitely generated group with an invariant component Δ_0 (cf. Gardiner's lecture).

An important <u>open question</u> (Ahlfors) is whether $\text{mes } \Lambda = 0$ for all finitely generated groups (cf. Maskit's lecture).

The inequality (7) is sharp. This is shown by the example of a <u>Schottky group</u> G of genus p constructed as follows. Let $C_1, C_1', C_2, \cdots, C_p'$ be $2p$ disjoint Jordan curves, each containing all others in the unbounded component of its complements and let $g_1, \cdots, g_p$ be Möbius transformations such that g_j maps the unbounded component of the complement of C_j onto the bounded component of the complement of C_j'. Then $g_1, \cdots, g_p$ are the free generators of a Kleinian group G. The limit set $\Lambda(G)$ is totally disconnected, $\text{mes } \Lambda = 0$, and Ω/G is of type $(p,0)$ so that $\text{Area } (\Omega/G) = 4\pi(p-1)$.

The classical <u>retrosection theorem</u> (suspected by Schottky before 1875, stated by Klein in 1882, proved by Koebe in 1907) asserts that every compact Riemann surface of genus $p > 0$ can be represented by a Schottky group.

(A <u>fundamental region</u> ω of a Kleinian group G is a set $\omega \subset \Omega$ such that mes $(\mathrm{Cl}(\omega)\setminus\omega) = 0$, no two interior points of ω are G equivalent, and every $z \in \Omega$ is G equivalent to some point of $\mathrm{Cl}(\omega)$. Clearly, $\Omega/G = \omega/G$. Example: for the Schottky group constructed above the region bounded by the $2p$ Jordan curves $C_1, \cdots, C'_p$ is a fundamental region.)

The best studied class of Kleinian groups is that of Fuchsian groups. A <u>Fuchsian group</u> G is a discrete subgroup of $\text{Möb}_{\mathbb{R}}$ (or conjugate to such a subgroup in Möb). It is called of the <u>first</u> or of the <u>second kind</u> according to whether Λ is the whole of $\hat{\mathbb{R}} = \mathbb{R} \cup \{\infty\}$ or a nowhere dense subset of $\hat{\mathbb{R}}$. In the first case, Ω has two components, U and the lower half-plane L, both invariant. In the second case, Ω is connected.

The most famous examples of Fuchsian groups of the first kind are the elliptic modular group $SL(2,\mathbb{Z})/\{\pm I\}$ and its subgroups of finite index.

The classical <u>limit circle theorem</u> (conjectured by Klein and by Poincaré in 1882, proved by Poincaré and by Koebe in 1907) asserts that, but for a few exceptional cases, every given Riemann surface with ramification points can be represented as U/G, G a Fuchsian group.

If G is a finitely generated Fuchsian group of the first kind, the Riemann surfaces U/G and L/G are mirror images of each other. In particular, they have the same signatures. Thus the inequality (8) is sharp.

(A fundamental region for a Fuchsian group G can be always chosen as a convex non-Euclidean polygon in U and its mirror image in L.)

A quasi-Fuchsian group G is a Kleinian group leaving a directed Jordan curve C on $\mathbb{C}$ fixed. G is called of the first kind, if $\Lambda = C$. The theorem on simultaneous uniformization [4] asserts, among other things, that any two given Riemann surfaces of the same signature can be represented by a given quasi-Fuchsian group. The only known proof of this theorem depends on the theory of quasiconformal mappings (cf. Earle's lecture).

A finitely generated Kleinian group G is called a b-group if it has a simply connected invariant component Δ_0. If so, there is a conformal bijection $W: L \to \Delta_0$ and a Fuchsian group Γ such that $G = W\Gamma W^{-1}$. G is called a boundary group if there is a sequence of conformal bijections $W_j: L \to W_j(L)$ such that $\lim W_j = W$ uniformly on compact subsets of L, and all groups $W_j \Gamma W_j^{-1} = G_j$ are quasi-Fuchsian. It is conjectured that all b-groups are boundary groups, hence the name.

It turns out that "almost all" boundary groups are <u>totally degenerate</u>, that is satisfy $\Delta_0 = \Omega$ (Bers and Maskit [8]). Indeed for any given Riemann surface S with ramification points, of finite type, and with a signature which makes the right side of (6) positive, there are uncountably many non-conjugate (in Möb) totally degenerate b-groups G with $\Omega/G = S$ [6]. But not a single such group has been constructed.

A b-group is called <u>regular</u> if it satisfies (8) with equality. All such groups have been constructed (Maskit [14]), and all of them are boundary groups (Marden, Harvey, Abikoff, to appear). A regular b-group represents a Riemann surface S and one or more surfaces $S_1, S_2, \cdots, S_\ell$ which may be thought to have been obtained by "drawing Jordan curves $C_1, \cdots, C_r$ on the mirror image $\bar{S}$ of S, not homotopic to a point on $\bar{S}\setminus\{\text{set of ramification points}\}$, and then contracting each of those curves into a puncture." (An exceptional case occurs if a C_j bounds a disc containing two ramification points of order 2.)

An intermediate type of a b-group is a <u>partially degenerate group</u>, with $\Omega \neq \Delta_0$ and (8) holding with a strict inequality. Such groups can be constructed [14] assuming the existence of totally degenerated b-groups.

The constructions mentioned above use Klein's combination theorem and its extensions by Maskit [12], [13], [15]. Here is another application.

Let $G_1, \cdots, G_r$ be a Fuchsian group acting on discs $\Delta_1, \cdots, \Delta_r$ with Δ_j/G_j of signature $(p_j, n_j; \nu, \ldots, \nu)$ with $n_j \geq 1$, $2 < \nu < \infty$ and $n_1 + \ldots + n_r = 2k$. Let G be the group generated by $G_1, \ldots, G_r$. If the Δ_j are sufficiently far apart, then G is Kleinian, has $r + 1$ non-conjugate components $\Delta_0, \Delta_1, \ldots, \Delta_r$, with Δ_0 invariant, and $S_0 = \Delta_0/G$ has signature $(p, k; \nu, \ldots, \nu)$ with $p = p_1 + \ldots + p_r$. (This follows from Klein's combination theorem.)

Now we divide the 2k ramification points on $\Delta_1/G_1, \ldots, \Delta_j/G_j$ into k pairs $(P_1, P_1') \cdots (P_r, P_r')$ such that, if one identifies the two points of a pair, Ω/G becomes $S_0 + \Sigma_0$, Σ_0 connected. This Σ_0 is a so-called "Riemann surface with k nodes."

For $i = 1, \ldots, k$, let Γ_i and Γ_i' be elliptic subgroups of $G_{j(i)}$ and $G_{\ell(i)}$ corresponding to P_i and P_i', and if $s_i \in C$, $|s_i|$ positive and small, let g_i be the (unique) loxodromic Möbius transformations, with $(\text{trace } g_i)^2 = 2 + s_i + s_i^{-1}$, with $g_i \Gamma_i g_i^{-1} = \Gamma_i'$, and with fixed points in $\Delta_{j(i)}$ and in $\Delta_{\ell(i)}$. If $s_i = 0$, let $g_i = \text{id}$.

Let $s = (s_1, \ldots, s_k) \in \mathbb{C}^k$ be a vector with small norm and let G_r be the group generated by G_0 and $g_1, \ldots, g_k$.

Then G_s is a Kleinian group (this follows by Maskit's second combination theorem). Also, $\Omega(G_s)/G_s$, with P_i identified to P_i' for $s_i = 0$, is $S_0 + \Sigma_s$ where Σ_s is a Riemann surface with $k - k(s)$ nodes, $k(s)$ being the number of nonvanishing components s_i of s.

The groups G_s can be used, instead of regular b-groups, in the theory of moduli of Riemann surfaces degenerating to a surface with nodes (cf. Abikoff's lecture).

An important class of Kleinian groups are web groups (Abikoff [2]), that is finitely generated Kleinian groups such that the stabilizer of each component is quasi-Fuchsian of the first kind.

We give only one, highly pathological, example. Let G be a Fuchsian group leaving the unit disc fixed and representing two compact surfaces of type $(p,0)$. Let G' be the group obtained from G by conjugating it by $z \mapsto \lambda z$, λ a large positive number. The group G^0 generated by G and G' is Kleinian and represents three compact surfaces of types $(p,0)$, $(p,0)$ and $(2p,0)$, respectively. (This follows from Klein's combination theorem.) Let α and $t \geq 0$ be real numbers, let G_t be the group obtained by conjugating G by $z \mapsto (1 + t)e^{i\alpha}z$, and let G^t be the group generated by G_t and G'. Maskit (to appear) showed that, for almost all α, there is a number $s > 0$ such that G^t is Kleinian for $0 \leq t \leq s$, G^t represents three surfaces of types $(p,0)$, (p,0) and (2p,0) for $0 \leq t \leq s$, and G^s is a web group representing only two surfaces, both of type (p,0).

In this example, just like in the case of partially and totally degenerate b-groups, a Riemann surface "disappeared." Its debris is, in some sense, hidden in the limit set. Could it be that such a limit set has positive measure?

REFERENCES

[1] Abikoff, W., Some remarks on Kleinian groups, Advances in the theory of Riemann surfaces, Ann. of Math. Studies, 66(1971), 1-5.

[2] __________, Residual limit sets of Kleinian groups, Acta Math., 130(1973), 127-144.

[3] Ahlfors, L. V., Finitely generated Kleinian groups, Amer. J. Math., 86(1964), 413-429; 87(1965), 759.

[4] Bers, L., Simultaneous uniformization, Bull. Amer. Math. Soc., 66(1960), 94-97.

[5] ________, Inequalities for finitely generated Kleinian groups, J. Analyse Math., 18(1967), 23-41.

[6] ________, On boundaries of Teichmüller spaces and on Kleinian groups I, Ann. of Math., 91(1970), 570-600.

[7] ________, Uniformization, moduli, and Kleinian groups, Bull. London Math. Soc., 4(1972), 257-300.

[8] Bers, L. and B. Maskit, On a class of Kleinian groups, in Contemporary Problems in the Theory of Analytic Functions, Nauka, Moscow, (1966), 44-47.

[9] Ford, L. R., Automorphic Functions, 2nd Ed. (Chelsea, New York, 1951).

[10] Griffiths, P. A., Complex analytic properties of certain Zariski open sets on algebraic varieties, Ann. of Math., 94(1971), 21-51.

[11] Kra, I., Automorphic Forms and Kleinian Groups, W. A. Benjamin, Reading, Massachusetts (1972).

[12] Maskit, B., On Klein's combination theorem I, Trans. Amer. Math. Soc., 120(1965), 499-509.

[13] __________, On Klein's combination theorem II, Trans. Amer. Math. Soc., 131(1968), 32-39.

[14] __________, On boundaries of Teichmüller spaces and on Kleinian groups II, Ann. of Math., 91(1970), 608-638.

[15] Maskit, B., On Klein's combination theorem III, Advances in the theory of Riemann surfaces, Ann. of Math. Studies, 66(1971), 297-316.

2. QUASICONFORMAL MAPPINGS AND UNIFORMIZATION

C.J. Earle*
Cornell University

Since the pioneering work of Teichmüller, quasiconformal maps have played a significant role in the study of Riemann surfaces and Fuchsian and Kleinian groups. In this brief talk I want to survey several aspects of the theory and indicate some applications to Kleinian groups.

§1. QUASICONFORMAL MAPS

1.1 Let D and D' be domains in $\mathbb{C}$ and $f:D \rightarrow D'$ a sense-preserving homeomorphism. For each z in D set

$$H(x) = \limsup_{r \to 0} \frac{L(z,r)}{\ell(z,r)}$$

where

$$L(z,r) = \max\{|f(\zeta) - f(z)|;|\zeta - z| = r\}$$
$$\ell\ z,r) = \min\{|f(\zeta) - f(z)|;|\zeta - z| = r\}.$$

we say that f is <u>quasiconformal</u> (qc) in D if and only if $H(z)$ is a bounded function on D (a fortiori, $H(z)$ is finite for every z in D). The qc mapping f is K-quasiconformal (K-qc) in D if and only if $H(z) \leq K$ for almost all z in

*The author thanks the National Science Foundation for financial support through Grant GP-28251.

D. Obviously any qc mapping is K-qc for some (finite) K.

1.2 Suppose $f: D \to D'$ is a sense-preserving C^1 diffeomorphism. Let f_z and $f_{\bar{z}}$ be the complex derivatives

$$f_z = \frac{1}{2}\left(\frac{\partial f}{\partial x} - i\frac{\partial f}{\partial y}\right),$$

$$f_{\bar{z}} = \frac{1}{2}\left(\frac{\partial f}{\partial x} + i\frac{\partial f}{\partial y}\right).$$

Since the Jacobian $|f_z|^2 - |f_{\bar{z}}|^2$ of f is positive, we see that

$$L(z,r) = r(|f_z(z)| + |f_{\bar{z}}(z)|) + o(r),$$

$$\ell(z,r) = r(|f_z(z)| - |f_{\bar{z}}(z)|) + o(r),$$

$$H(z) = \frac{|f_z(z)| + |f_{\bar{z}}(z)|}{|f_z(z)| - |f_{\bar{z}}(z)|}.$$

Therefore the C^1 diffeomorphism f is K-qc if and only if

$$(1) \qquad |f_{\bar{z}}(z)| \leq \frac{K-1}{K+1}|f_z(z)|$$

for all z in D.

1.3 We want to extend the criterion (1) for quasiconformality to the general case. We need to recall the definition of generalized (distribution) derivatives.

We say that f has generalized derivatives f_z, $f_{\bar{z}}$ in D if and only if f_z and $f_{\bar{z}}$ are locally L^2 functions in D satisfying

$$0 = \iint (\varphi f_z + f\varphi_z)dxdy = \iint (\varphi f_{\bar z} + f\varphi_{\bar z})dxdy$$

for all smooth functions φ with compact support in D. The <u>analytic definition</u> for quasiconformal maps states that the homeomorphism $f: D \to D'$ is K-qc if and only if f has generalized derivatives in D satisfying (1) almost everywhere. The equivalence of this definition with our first one is proved, for instance, in Lehto - Virtanen [5], §4 of Chapter 4. Notice that 1-qc maps are conformal, by Weyl's lemma.

1.4 We list some useful properties of qc mappings.

<u>Proposition</u> (see [1], [2], or [5]). <u>Let</u> $f: D \to D'$ <u>be</u> K-qc. <u>Then</u>

(a) f <u>is differentiable</u> a.e.

(b) $|f_z| > 0$ a.e.

(c) $\operatorname{mes}(f(E)) = \iint_E (|f_z|^2 - |f_{\bar z}|^2)dxdy$

<u>for all measurable sets</u> $E \subset D$.

(d) $f^{-1}: D' \to D$ <u>is</u> K-qc. <u>If</u> $g: D' \to D''$ <u>is</u> K'-qc, <u>then</u> $g \circ f$ is KK'-qc <u>in</u> D.

§2. BELTRAMI EQUATIONS

2.1 If f is qc in D, then f solves the <u>Beltrami equation</u>

$$f_{\bar z} = \mu f_z \tag{2}$$

in D, where $\mu(z) = f_{\bar{z}}(z)/f_z(z)$ is a measurable function whose L^∞ norm in D is less than one. Conversely, for any such μ there is a qc map in D which solves (2). Further, this map can be chosen to depend nicely on the parameter μ.

To be more explicit, let $M(\mathbb{C})$ be the open unit ball in the Banach space $L^\infty(\mathbb{C},\mathbb{C})$.

Theorem (Ahlfors-Bers [2]): *For each μ in $M(\mathbb{C})$ there is a unique qc map w^μ of C onto itself that fixes zero and one and solves the equation (2). For any fixed ζ in $\mathbb{C}$ the map*

$$\mu \longmapsto w^\mu(\zeta)$$

is a holomorphic function on $M(\mathbb{C})$. Explicitely we have

$$w^\mu(\zeta) = \zeta + P\mu(\zeta) + o(\|\mu\|), \quad \mu \to 0,$$

where $o(\|\mu\|)$ is uniform on compact subsets of $\mathbb{C}$ and $P\mu$ *is given by*

$$P\mu(\zeta) = -\frac{1}{\pi}\iint \frac{\zeta(\zeta-1)}{z(z-1)}\frac{\mu(z)}{(z-\zeta)}\,dxdy.$$

2.2 For any domain D, let M(D) be the set of μ in $M(\mathbb{C})$ which vanish a.e. in the complement of D. If $\mu \in M(D)$, w^μ is a qc self-map of $\mathbb{C}$ which is conformal in the exterior of D.

If $\mu \in M(U)$ we denote by w_μ the unique qc self-map of the upper half-plane U that fixes 0,1, and ∞ and solves (2) in U. Notice that $w_\mu = w^\nu$, where $\nu \in M(\mathbb{C})$ is defined by

(3) $\nu(z) = \mu(z)$, $\nu(\bar{z}) = \bar{\mu}(z)$ for all $z \in U$.

Further, $w^{\mu} \circ (w_{\mu})^{-1}$ is the conformal map of U onto $w^{\mu}(U)$ that fixes $0, 1$, and ∞.

§3. EXTREMAL SELF-MAPS OF U

3.1 Qc maps have the following important compactness property. Let (μ_n) be any sequence in $M(\mathbb{C})$ with

$$\|\mu_n\| \leq k < 1 \quad \text{for all } n.$$

Let $w_n = w^{\mu_n}$. Then a subsequence w_{n_j} converges uniformly on compact sets to a qc map w^{ν}, and

$$\|\nu\| \leq \liminf_{j \to \infty} \|\mu_{n_j}\|.$$

3.2 We call the qc map $w_{\mu}: U \to U$ extremal if and only if $\|\mu\| \leq \|\nu\|$ for all ν in $M(U)$ with $w_{\mu} = w_{\nu}$ on the real axis. Given any w_{ν}, the compactness property implies that there is an extremal w_{μ} with the same boundary values. It is of considerable importance to describe the extremal maps w_{μ} in terms of their Beltrami coefficients μ.

R.S. Hamilton [4] gave an analytic condition satisfied by μ if w_{μ} is extremal. To formulate this condition we introduce the Banach space $A(U)$ of L^1 holomorphic functions in U. Its annihilator $A(U)^{\perp}$ consists of the L^{∞} functions $\sigma(z)$ on U such that

$$0 = \iint_U \varphi(z)\sigma(z)dxdy \quad \text{for all} \quad \varphi \in A(U).$$

μ in M(U satisfies Hamilton's condition if and only if

(4) $$\|\mu\| \leq \|\mu + \sigma\| \quad \text{for all} \quad \sigma \in A(U)^{\perp}.$$

Recently, Reich and Strebel [6] have proved that w_μ is extremal if and only if μ satisfies Hamilton's condition.

3.3 We call w_μ a Teichmüller mapping and μ a Teichmüller differential if and only if

$$\mu = k\overline{\varphi}/|\varphi|$$

where $0 \leq k < 1$ and φ is a holomorphic function in U. It follows from Reich and Strebel [6], or by direct arguments (Strebel [7]) that w_μ is extremal if $\varphi \in A(U)$. It is an important unsolved problem to give necessary and sufficient conditions on φ to make w_μ extremal.

§4. BELTRAMI COEFFICIENTS FOR A GROUP

4.1 For any Kleinian group G and any w^μ, the group

$$G^\mu = w^\mu G(w^\mu)^{-1} = \{w^\mu \circ g \circ (w^\mu)^{-1};\ g \in G\}$$

is a group of homeomorphisms of $\hat{\mathbb{C}}$, acting discontinuously on $w^\mu(\Omega)(G))$. In order that G^μ be a Kleinian group, it is necessary and sufficient to have $w^\mu \circ g \circ (w^\mu)^{-1}$ conformal for all g in G. This happens if and only if

$$(5) \qquad (\mu \circ g)\overline{g'}/g' = \mu \quad \text{for all} \quad g \in G.$$

Let Σ be any invariant union of components of Ω. We denote by $M(G,\Sigma)$ the set of μ in $M(\Sigma)$ satisfying (5). μ in $M(G,\Sigma)$ is called a <u>Beltrami coefficient for</u> G <u>with support in</u> Σ. For each such μ, the group G^μ is Kleinian. Theorem 2.2 implies easily that the elements of G^μ depend holomorphically on the parameter μ. The groups G^μ are called qc <u>deformations</u> of G. They play a central role in the study of variations of the group G.

4.2 If G is a Fuchsian group on U, we denote by $M(G)$ the set of μ in $M(U)$ satisfying (5). If $\mu \in M(G)$, then ν defined by (3) belongs to $M(G,\Omega)$, so $G_\mu = w_\mu G(w_\mu)^{-1}$ is a Fuchsian group. Since ν depends real analytically on μ, the elements of G_μ depend real analytically on μ, and we obtain the family of qc Fushsian deformations G_μ of G.

For most μ in $M(G)$ the Kleinian group G^μ will not be Fushsian. However, every element of G^μ will map the sets $w^\mu(U)$ and $w^\mu(\hat{\mathbb{R}})$ onto themselves. These groups G^μ are the <u>quasi-Fuchsian</u> groups.

4.3 This construction of quasi-Fuchsian groups, due to Bers, leads easily to a proof of the theorem on <u>simultaneous uniformization</u> mentioned in the first lecture.

Let S_1 and S_2 be closed Reimann surfaces of genus $g \geq 2$. Choose a Fuchsian group G so that U/G is S_1^*, the mirror image of S_1. Then of course, S_1 is the quotient of the lower half-plane L by G. Choose another Fuchsian group H so that U/H is S_2. Let f be a sense-preserving diffeomorphism of S_1^* onto S_2, and let $w: U \to U$ be a lifting of f. Then w is qc, $\mu = w_{\bar{z}}/w_z$ belongs to $M(G)$, and G^μ is a quasi-Fuchsian group. Obviously $w^\mu(U)/G^\mu$ is S_2, and $w^\mu(L)/G^\mu$ is S_1.

§5. EXTREMAL MAPS FOR A GROUP

Consider a Fuchsian group G on U. If $\mu \in M(G)$ we call w_μ extremal for G if and only if $\|\mu\| \leq \|\nu\|$ for all $\nu \in M(G)$ such that $w_\mu = w_\nu$ on $\mathbb{R}$. Again, compactness properties ensure that extremal maps exist, and it is important to describe them. Hamilton [4] proved that if w_μ is extremal for G then μ satisfies an appropriate analog of (4). The converse has recently been established by Strebel [8] for finitely generated G. If U/G is compact (or has finite noneuclidean area), Teichmüller's theorem [3] tells us that w_μ is extremal for G if and only if

$$\mu = k\bar{\varphi}/|\varphi|, \quad 0 \leq k < 1,$$

where φ is a quadratic differential for G. That is, φ is holomorphic in U and satisfies

$$\varphi(g(z))g'(z)^2 = \varphi(z), \quad \text{all } g \in G \text{ and } z \in U.$$

REFERENCES

[1] L.V. Ahlfors, *Lectures on Quasiconformal Mappings*, Van Nostrand-Reinhold, Princeton, New Jersey, 1966.

[2] L.V. Ahlfors and L. Bers, Riemann's mapping theorem for variable matrices, *Ann. of Math.*, 72(1960), 385-404.

[3] L. Bers, Quasiconformal mappings and Teichmüller's theorem, in *Analytic Functions*, pp. 89-119, Princeton University Press, Princeton, New Jersey, 1960.

[4] R.S. Hamilton, Extremal quasiconformal mappings with prescribed boundary values, *Trans. Amer. Math. Soc.*, 138(1969), 399-406.

[5] O. Lehto and K.I. Virtanen, *Quasikonforme Abbildungen*, Springer-Verlag, Berling, 1965.

[6] E. Reich and K. Strebel, Extremal quasiconformal mappings with given boundary values, to appear.

[7] K. Strebel, Zur Frange der Eindentigkeit extremaler quasiknoformer Abbildungen des Einheitskreises II, *Comment. Math. Helv.*, 39(1964), 77-89.

[8] K. Strebel, On the trajectory structure of quadratic differentials, in *Discontinuous Groups and Reimann Surfaces*, pp. 419-438, Princeton University Press, Princeton, New Jersey, 1974.

3. AUTOMORPHIC FORMS AND EICHLER COHOMOLOGY

Frederick P. Gardiner
Boston College

One method of obtaining information about a Kleinian group is to study the properties of its automorphic forms and Eichler cohomology groups. This chapter shows how this method leads to Ahlfors' finiteness theorem and Bers' area theorems.

A good listing of the articles and research done in related areas can be found in the bibliography to [6].

§1. AUTOMORPHIC FORMS AND POINCARÉ THETA SERIES

In this section we assume that Γ is a nonelementary Kleinian group acting discontinuously on a connected open subset Δ of $\mathbb{C}$. In applications, Δ will be an invariant component of $\Omega(\Gamma)$, the set of discontinuity of Γ. Under these conditions, Δ is a hyperbolic Riemann surface and has a Poincaré metric $\lambda = \lambda_\Delta$.

If f is a holomorphic mapping from D_1 to D_2 and φ is a function defined on D_2, then $f^*_{p,q}\varphi(z) = \varphi(f(z))f'(z)^p\overline{f'(z)}^q$ is a function defined on D_1. In this definition p and q must be half-integers and $p + q$ an integer. We also use the convention that $f^*_p = f^*_{p,o}$.

With this notation the invariance property of λ can be expressed by $f^*_{1/2,1/2}\lambda_{f(\Delta)} = \lambda_\Delta$ for any conformal mapping f.

A measurable automorphic form of weight $(-2q)$ in Δ is a function μ with the property that $\gamma_q^*\mu = \mu$ for all $\gamma \in \Gamma$. We define two Banach spaces of such forms for any integer $q \geq 2$. $L_q^\infty(\Delta, \Gamma)$ = all measurable functions μ on Δ such that $\gamma_q^*\mu = \mu$ and $\|\mu\|_\infty < \infty$ where

$$\|\mu\|_\infty = \sup_z \{\lambda^{-q}(z)|\mu(z)|\}.$$

$L_q^1(\Delta, \Gamma)$ = all measurable functions ν on Δ such that $\gamma_q^*\nu = \nu$ and $\|\nu\|_\Gamma < \infty$ where

$$\|\nu\|_\Gamma = \iint_\omega \lambda^{2-q}\,|\nu|\,|dz \wedge \overline{dz}|$$

and ω is any fundamental domain for Γ in Δ.

We shall require that all definitions of spaces that we make be invariant under f_q^* whenever f is a conformal mapping.

In particular, if $\infty \notin \Delta$, we define $B_q(\Delta, \Gamma)$ and $A_q(\Delta, \Gamma)$ to be the closed subspaces of holomorphic functions in $L_q^\infty(\Delta, \Gamma)$ and $L_q^1(\Delta, \Gamma)$, respectively. When $\infty \in \Delta$, the condition that a function φ is holomorphic at ∞ translates into the condition that $\varphi(z) = 0(|z|^{-2q})$ as $z \to \infty$.

<u>Remark.</u> In the case that Γ is the trivial group we write $B_q(\Delta)$ and $A_q(\Delta)$ instead of $B_q(\Delta, \Gamma)$ and $A_q(\Delta, \Gamma)$. If we let $\Delta = U$, the space $A_2(U)$ coincides with the space $A(U)$ introduced by Earle in Lecture 2.

The definitions of these spaces generalize to the case where

Δ is replaced by Σ, an invariant union of components of $\Omega(\Gamma)$. λ_Σ is defined by stipulating that its restriction to any component Δ of Σ be equal to λ_Δ. Then the spaces $L_q^1(\Sigma, \Gamma)$, $L_q^\infty(\Sigma, \Gamma)$ and $A_q(\Sigma, \Gamma)$, $B_q(\Sigma, \Gamma)$ are defined just as before.

There is a pairing between $L_q^1(\Sigma, \Gamma)$ and $L_q^\infty(\Sigma, \Gamma)$ given by the Petersson scalar product:

$$(\mu, \nu) = \iint_\omega \lambda^{2-2q} \mu \bar{\nu} \, |dz \wedge d\bar{z}|,$$

where ω is a fundamental domain for Γ in Σ. By standard methods in analysis one can show that this pairing establishes an antilinear isomorphism between the dual of $L_q^1(\Sigma, \Gamma)$ and $L_q^\infty(\Sigma, \Gamma)$.

We will need the following theorems about automorphic forms and theta series.

Theorem 1. (Bers [5]) The Petersson scalar product establishes an antilinear isomorphism between $B_q(\Sigma, \Gamma)$ and the dual space to $A_q(\Sigma, \Gamma)$.

Definition. Let φ be a holomorphic function in Σ. The Poincaré theta series is defined by

$$\Theta\varphi(z) = \sum_{\gamma \in \Gamma} \varphi(\gamma(z))\gamma'(z)^q, \qquad z \in \Sigma$$

whenever the right side converges absolutely and uniformly on compact subsets of Σ.

Theorem 2. $\Theta : A_q(\Sigma) \longrightarrow A_q(\Sigma, \Gamma)$ *is a norm decreasing, surjective linear mapping. To each* $\psi \in A_q(\Sigma, \Gamma)$ *there corresponds a* $\varphi \in A_q(\Sigma)$ *such that* $\Theta\varphi = \psi$ *and* $\|\varphi\| \leq \frac{2q-1}{q-1} \|\psi\|_\Gamma$.

Definition. Let $S = \Delta/\Gamma$ be of finite type. Let $\hat{S}$ be S with the parabolic punctures restored. For each $p \in \hat{S}$, let $\nu(p) = \infty$ if p is a puncture, the order of p if p is a ramification point, and 1 otherwise.

The space of cusp forms $C_q(\Delta, \Gamma)$ is the space of holomorphic automorphic q-forms φ in Δ such that, when φ is viewed as a differential on $\hat{S}$, φ has a pole at p of order at most $[q - \frac{q}{\nu(p)}]$. (Here, as usual, $[x]$ = the largest integer $\leq x$ and we use the convention that $[q - \frac{q}{\infty}] = q - 1$. Notice that $\nu(p) = 1$ implies φ is holomorphic at p.)

Lemma 1. (Ahlfors [1], page 416) If $p \in \hat{S} - S$, then there exists $\gamma \in \Gamma$ and a Möbius transformation U such that $U^{-1}\gamma U(z) = z + 1$ and $U^{-1}(\Delta)$ contains the set $\{z \mid \operatorname{Im} z > c\}$ and such that $\pi(U(z))$ approaches p as $\operatorname{Im} z \to \infty$.

Theorem 3. *If* Δ/Γ *is of finite type, then* $A_q(\Delta, \Gamma) = B_q(\Delta, \Gamma) = C_q(\Delta, \Gamma)$ *and the dimension of these spaces is*

$$(2q-1)(g-1) + \sum_{p \in \hat{S}} [q - \frac{q}{\nu(p)}]$$

where $g = \text{genus}\,(\hat{S})$.

Lemma 2. Suppose Δ/Γ is of finite type. The following conditions on a holomorphic automorphic q-form φ in Δ are equivalent:

1) $\sup_{z\in\Delta} \{\lambda^{-q}(z) \mid \varphi(z)|\} < \infty$,

2) $\iint_{\omega} \lambda^{2-q} |\varphi| \, |dz \wedge d\bar{z}| < \infty$,

3) If $\lim_{n\to\infty} z_n = \zeta$ where z_n is contained in cusped region belonging to a puncture on $\hat{S}$ and $\zeta \in \Lambda$ then $\lim_{n\to\infty} \varphi(z_n) = 0$.

Complete proofs to these results are contained in Kra [6].

§2. EICHLER COHOMOLOGY

Let Π_{2q-2} be the vector space of polynomials of degree $\leq 2q-2$. If Γ is a group of Möbius transformations, then Γ acts on Π_{2q-2} on the right by defining $v \cdot \gamma = v(\gamma(z))\gamma'(z)^{1-q}$ for every $v \in \Pi_{2q-2}$ and every $\gamma \in \Gamma$. If $\gamma(z) = (az+b)(cz+d)^{-1}$ where $ad-bc = 1$, then $\gamma'(z) = (cz+d)^{-2}$ and, using this fact, it is easy to see that $v \cdot \gamma$ is in Π_{2q-2} whenever v is in Π_{2q-2}. Furthermore, by use of the chain rule, it is easy to check that $v \cdot (\gamma_1\gamma_2) = (v\cdot\gamma_1) \cdot \gamma_2$.

Under these circumstances, one can form the group cohomology groups $H^j(\Gamma, \Pi_{2q-2})$. We will only be concerned with the first cohomology group and now give an explicit description of it. A mapping $P : \Gamma \longrightarrow \Pi_{2q-2}$ is called a 1-cocycle if $P_{\gamma_1\gamma_2} = P_{\gamma_1} \cdot \gamma_2 + P_{\gamma_2}$ for all γ_1 and γ_2. A 1-coboundary is a mapping $P : \Gamma \longrightarrow \Pi_{2q-2}$ of the form $P_\gamma = v \cdot \gamma - v$. One easily checks that any 1-coboundary is a 1-cocycle. The first cohomology group $H^1(\Gamma, \Pi_{2q-2})$ is the vector space of 1-cocycles factored by the vector space of 1-coboundaries.

Let B be a Möbius transformation and $\hat{\Gamma} = B^{-1}\Gamma B$. Then conjugation by B^{-1} induces an isomorphism between $H^1(\Gamma, \Pi_{2q-2})$ and $H^1(\hat{\Gamma}, \Pi_{2q-2})$. The mapping is determined by sending the cocycle P into the cocycle $\hat{P}$ where $\hat{P}(B^{-1}\gamma B) = P(\gamma) \cdot B = B^*_{1-q}P(\gamma)$ for all $\gamma \in \Gamma$. It is easy to see that the mapping $P \longrightarrow \hat{P}$ is invertible and preserves coboundaries. It is important to realize that if Γ_1 and Γ_2 are Kleinian groups and $g : \Gamma_1 \longrightarrow \Gamma_2$ is an algebraic isomorphism, there will not, in general, be any relationship between $H^1(\Gamma_1, \Pi_{2q-2})$ and $H^1(\Gamma_2, \Pi_{2q-2})$. The structure of $H^1(\Gamma, \Pi_{2q-2})$ depends on the geometric manner in which Γ is a subgroup of the full Möbius group.

However, one can find a bound on $\dim H^1(\Gamma, \Pi_{2q-2})$ in terms of the number of generators of Γ.

Lemma 3. Suppose $q \geq 2$ and Γ is a nonelementary Kleinian group generated by N elements. Then $\dim H^1(\Gamma, \Pi_{2q-2}) \leq (2q-1)(N-1)$.

Proof: Let $v \in \Pi_{2q-2}$ and assume that $\delta v = v \cdot \gamma - v = 0$. This would mean that $v(\gamma(z))\gamma'(z)^{1-q} = v(z)$ for every γ in Γ. Thus if $v(z_0) = 0$ then $v(\gamma(z_0)) = 0$ for all γ and this implies v has infinitely many zeroes since Γ is nonelementary. Therefore, δ is injective and the space of coboundaries has dimension $2q-1$ (which is the dimension of Π_{2q-2}). Since a cocycle is uniquely determined by its values on generators of Γ, the dimension of the space of cocycles is $\leq (2q-1)N$. We conclude that $\dim H^1(\Gamma, \Pi_{2q-2}) \leq (2q-1)(N-1)$.

Next, we introduce an analytic way to obtain elements of $H^1(\Gamma, \Pi_{2q-2})$. As usual, we assume Γ is nonelementary (so that $\Lambda(\Gamma)$ is an infinite set). Let Σ be a union of components of Ω invariant under the action of Γ. By a <u>generalized Beltrami coefficient</u> for Γ with support in Σ we mean a measurable function μ, with $\mu \mid (\mathbb{C}-\Sigma) = 0$, such that $\gamma^*_{1-q,1}\mu = \mu$ for all $\gamma \in \Gamma$ and such that $|\mu(z)| \leq (\text{const.})\,\lambda^{2-q}(z)$. (For $q = 2$, these are the ordinary Beltrami coefficients discussed in Lecture 2.)

A continuous function $F(z)$ will be called a <u>potential</u> for the generalized Beltrami coefficient μ if

(2.1) $$\frac{\partial F}{\partial \bar{z}} = \mu \text{ in the sense of distribution theory}$$

and if

(2.2) $$F(z) = 0(|z|^{2q-2}), \quad z \to \infty.$$

If F is a potential for μ and if $v \in \Pi_{2q-2}$, then $F + v$ is also a potential for μ. Conversely, if F_1 and F_2 are potentials for μ, then $F_1 - F_2 \in \Pi_{2q-2}$.

<u>Remark.</u> If A is a Möbius transformation, μ a generalized Beltrami coefficient for Γ, and F a potential for μ, then $\hat{\mu} = A^*_{1-q,1}\mu$ is a generalized Beltrami coefficient for $A^{-1}\Gamma A$ and $\hat{F} = A^*_{1-q}F$ is a potential for $\hat{\mu}$.

Lemma 4. If μ is a generalized Beltrami coefficient for the nonelementary Kleinian group Γ with support in Σ, and if $a_1, a_2, \ldots, a_{2q-1}$ are distinct points in $\Lambda(\Gamma)$, then

$$(2.3) \qquad F(z) = \frac{(z-a_1)\cdots(z-a_{2q-1})}{2\pi i} \iint_{\Sigma} \frac{\mu(\zeta)d\zeta \wedge \overline{d\zeta}}{(\zeta-z)(\zeta-a_1)\cdots(\zeta-a_{2q-1})}$$

is a potential for μ.

Proof: Let A be a Möbius transformation and let $\hat{F} = A^*_{1-q}F$. One finds by change of variable that

$$\hat{F}(z) = \frac{(z-\hat{a}_1)\cdots(z-\hat{a}_{2q-1})}{2\pi i} \iint_{A^{-1}(\Sigma)} \frac{\hat{\mu}(\zeta)d\zeta \wedge \overline{d\zeta}}{(\zeta-z)(\zeta-\hat{a}_1)\cdots(\zeta-\hat{a}_{2q-1})}$$

where $\hat{\mu} = A^*_{1-q}\mu$ and $\hat{a}_j = A^{-1}(a_j)$, $j = 1, \ldots, 2q-1$. Using this observation and the remark above, we may assume without loss of generality that $\infty \in \Omega$ and Λ is bounded. It is simple consequence of Schwarz's lemma that if $\infty \in \Omega$, then $1/\lambda(z) = 0(|z|^2)$, (see Kra [6], page 168). Hence, from the fact that $|\mu| \leq (\text{const.})\,\lambda^{2-q}$, we conclude that $\mu(z) = 0(|z|^{2q-4})$, $z \to \infty$. It follows that the integrand in (2.3) is $0(|\zeta|^{-4})$ and the integral converges absolutely for all $z \neq a_j$. Moreover, the integral is $0(|\log|z-a_j||)$ as $z \to a_j$ and, therefore, $F(z) = 0(|z-a_j|\,|\log|z-a_j||)$ as $z \to a_j$. Thus by letting $F(a_j) = 0$, F becomes continuous at a_j.

We must show that F is continuous everywhere and satisfies (2.1) and (2.2). By elementary estimates one shows

$F(z) = O(|z|^{2q-2})$ as $z \to \infty$ and, by routine use of the change of variable formula, one can show that, for every $R > 0$ there is a constant $C(R)$ such that

$$|F(z) - F(w)| \leq C(R) \; |z-w| \, \big|\log|z-w|\big|$$

whenever $|z|$ and $|w| < R$. (See Kra [6], page 137-141.)

It remains to show that $\partial F/\partial \bar{z} = \hat{\mu}$ in the sense of distributions. Let φ be a test function, that is, a C^∞ function with compact support. We must show that $-\iint F\varphi_{\bar{z}} \, dz \wedge \overline{dz} = \iint \mu\varphi \, dz \wedge \overline{dz}$ for every test function φ. Let $p(z) = (z-a_1) \cdots (z-a_{2q-1})$. Then

$$-\iint F\varphi_{\bar{z}} \, dz \wedge \overline{dz} = -\iint \varphi_{\bar{z}} \frac{p(z)}{2\pi i} \iint \frac{\mu(\zeta)}{(\zeta - z)p(\zeta)} \, d\zeta \wedge \overline{d\zeta} \; dz \wedge \overline{dz}$$

$$= \frac{1}{2\pi i} \iint \frac{\mu(\zeta)}{p(\zeta)} \iint \frac{\bar{\partial}(\varphi(z)p(z))}{(z-\zeta)} \, dz \wedge \overline{dz} \; d\zeta \wedge \overline{d\zeta}$$

by Fubini's theorem. By the general Cauchy integral formula, the inside integral in this last expression yields $2\pi i \, \varphi(\zeta)p(\zeta)$, and the lemma is proved.

Let $M_q(\Sigma, \Gamma)$ be the space of all generalized Beltrami coefficients with support in Σ.

Theorem 4. _There is a canonical linear mapping_

$$\beta : M_q(\Sigma, \Gamma) \longrightarrow H^1(\Gamma, \Pi_{2q-2}).$$

Proof: Given any $\mu \in M_q(\Sigma, \Gamma)$, form a potential F for μ. For each $\gamma \in \Gamma$ let $P_\gamma(z) = F(\gamma(z))\gamma'(z)^{1-q} - F(z)$. From $F(z) = 0(|z|^{2q-2})$, one easily sees that $P_\gamma(z) = 0(|z|^{2q-2})$. Using the facts that $\gamma^*_{1-q,1}\mu = \mu$ and $\frac{\partial F}{\partial \bar{z}} = \mu$, one sees that $\frac{\partial}{\partial \bar{z}} P_\gamma(z) = 0$ for all $z \in \mathbb{C}$. Therefore $P_\gamma \in \prod_{2q-2}$. It is easy to show that P_γ satisfies the cocycle relation. The choice of a potential F for μ is not unique, but obviously the cohomology class of P is uniquely determined by μ. It is equally clear that the mapping $\beta(\mu)$ = (the cohomology class of P) is linear.

The next lemma gives us a useful condition for deciding when the cohomology class associated to μ is zero.

Lemma 5. Let $\mu \in M_q(\Sigma, \Gamma)$. The following conditions are equivalent:

(i) $\beta(\mu) = 0$.

(ii) μ has a potential F such that $\gamma^*_{1-q}F = F$ for all $\gamma \in \Gamma$.

(iii) μ has a potential F such that $F \mid \Lambda = 0$.

Proof: If (i) holds, then μ has a potential F_0 such that $\gamma^*_{1-q}F_0 - F_0 = \gamma^*_{1-q}v_0 - v_0$ for some $v_0 \in \prod_{2q-2}$ and for every $\gamma \in \Gamma$. Thus $F = F_0 - v_0$ has property (ii). If (ii) holds, then μ has a potential F such that $F(\zeta)\gamma'(\zeta)^{1-q} = F(\zeta)$ for every loxodromic (or hyperbolic) fixed point $\zeta \in \Lambda$. But such fixed points are dense in Λ and if γ is the loxodromic transformation with fixed point ζ, then $\gamma'(\zeta) \neq 1$. One concludes, by the continuity of F, that $F \mid \Lambda = 0$. If (iii) holds, then each polynomial $P_\gamma(z) = F(\gamma(z))\gamma'(z)^{1-q} - F(z)$ vanishes on Λ, an infinite set, and therefore $P_\gamma(z) = 0$.

Recall that the elements of $B_q(\Gamma, \Sigma)$ are holomorphic functions ψ with support in Σ such that $\psi(\gamma(z))\gamma'(z)^q = \psi(z)$ and such that $\lambda^{-q}(z)|\psi(z)| \leq (\text{const.})$. Therefore, if we let $\mu = \lambda^{2-2q}\overline{\psi}$, it is obvious that $|\mu(z)| \leq (\text{const.})\lambda^{2-q}(z)$. Furthermore, one can easily show that $\gamma^*_{1-q,1}\mu = \mu$, and so $\mu \in M_q(\Sigma, \Gamma)$.

<u>Theorem 5.</u> <u>Consider the mappings</u>

$$B_q(\Sigma,\Gamma) \xrightarrow{i} M_q(\Sigma,\Gamma) \xrightarrow{\beta} H^1(\Gamma, \Pi_{2q-2})$$

<u>where</u> $i(\psi) = \lambda^{2-q}\overline{\psi}$ <u>and</u> β <u>is defined as before. Then</u> $\mathrm{im}(\beta \circ i) = \mathrm{im}\,\beta$.

<u>Proof:</u> What we must show is that the cohomology class of any generalized Beltrami coefficient is the same as the cohomology class of $i(\psi)$ for some $\psi \in B_q(\Sigma, \Gamma)$. Note that if $\{a_1, \ldots, a_{2q-1}\}$ are 2q-1 distinct points in Λ and $\infty \in \Omega$, then

$$\Phi^z(\zeta) = \frac{1}{2\pi i}\,\frac{(z-a_1)\cdots(z-a_{2q-1})}{(\zeta-z)(\zeta-a_1)\cdots(\zeta-a_{2q-1})}$$

is an element of $A_q(\Omega)$ for each $z \in \Lambda - \{a_1, \ldots, a_{2q-1}\}$. By theorem 2, we can form $\varphi^z(\zeta) = \Theta_q\Phi^z(\zeta)$. Let μ be a generalized Beltrami coefficient for Γ with support in Σ. A potential F for μ is given by

$$\text{(2.4)} \qquad F(z) = \iint_{\mathbb{C}} \Phi^z(\zeta)\mu(\zeta)d\zeta \wedge \overline{d\zeta}$$

$$= \iint_{\Sigma} \Phi^z(\zeta)\mu(\zeta)d\zeta \wedge \overline{d\zeta}$$

By using the invariance properties of μ and the fact that $\Theta_q\Phi^z = \varphi^z$, one can show that

(2.5) $$F(z) = \iint_{\Sigma/\Gamma} \varphi^z(\zeta)\mu(\zeta)d\zeta \wedge \overline{d\zeta}.$$

Since the restriction of φ^z to Σ is an element of $A_q(\Sigma,\Gamma)$, by the duality between A_q and B_q, we know there exists $\psi \in B_q(\Sigma,\Gamma)$ such that

$$F(z) = \iint_{\Sigma/\Gamma} \varphi^z(\zeta)\lambda^{2-2q}(\zeta)\overline{\psi}(\zeta)d\zeta \wedge \overline{d\zeta}.$$

This shows that F is also a potential for the generalized Beltrami coefficient $i(\psi)$ and completes the proof of the theorem.

We now raise a crucial question: is $\beta \circ i$ an injective mapping? This amounts to asking whether the cohomology class of the potential function F is uniquely determined by the automorphic form $\psi \in B_q(\Sigma,\Gamma)$. As a step towards giving an answer, we prove

Theorem 6. If the functions $\Phi^z(\zeta)$, as z varies over the set $\Lambda - \{a_1, \ldots, a_{2q-1}\}$, span a dense subspace of $A_q(\Sigma)$, then $\beta \circ i : B_q(\Sigma,\Gamma) \longrightarrow H^1(\Gamma, \prod_{2q-2})$ is injective.

Proof: Suppose $\beta \circ i(\psi) = 0$. Then by lemma 4, there is a potential F for $\lambda^{2-2q}\overline{\psi}$ such that $F \mid \Lambda = 0$. In fact, F is given by

$$F(z) = \iint_{\Sigma} \Phi^z(\zeta)\lambda^{2-2q}(\zeta)\overline{\psi}(\zeta)d\zeta \wedge \overline{d\zeta}$$

since both the integral and F are potentials for $\lambda^{2-2q}\overline{\psi}$ which vanish at the $2q-1$ points, $a_1, \ldots, a_{2q-1}$. Since by hypothesis $F(z) = 0$ for all $z \in \Lambda$, the fact that $\psi = 0$ follows from the assumption that the functions Φ^z span a dense subspace of $A_q(\Sigma)$ and from the fact that the Petersson scalar product is a nondegenerate pairing between $A_q(\Sigma)$ and $B_q(\Sigma)$.

Remarks 1. Ultimately, we will show $\beta \circ i$ is injective when $q = 2$ and $\beta \circ i$ is injective for Γ finitely generated and $q \geq 2$. It is not known whether $\beta \circ i$ is injective when Γ is infinitely generated and $q > 2$.

2. By introducing the space of Eichler integrals, Kra has shown that $H^1(\Gamma, \prod_{2q-2})$ is the middle term of an exact sequence and that it is the direct sum of the space $B_q(\Sigma, \Gamma)$ and an appropriate space of Eichler integrals. For an exposition of this theory and a listing of Kra's papers on this subject, see [6].

§3. AN APPROXIMATION THEOREM

In this section we will concentrate on the case $q = 2$. As before, let

$$\Phi^z(\zeta) = \frac{1}{2\pi i} \frac{(z-a_1)(z-a_2)(z-a_3)}{(\zeta-z)(\zeta-a_1)(\zeta-a_2)(\zeta-a_3)}$$

where a_1, a_2, a_3 are three distinct fixed points in Λ and z varies over the set $\Lambda - \{a_1, a_2, a_3\}$.

We have seen that in order to prove that $\beta \circ i$ is injective when $q = 2$, it suffices to prove the following theorem.

<u>Theorem 7.</u> (Bers [2]) <u>Let</u> Λ <u>be the limit set of a nonelementary Kleinian group</u> Γ, (Γ <u>may be infinitely generated</u>). <u>Then the functions</u> $\Phi^z(\zeta)$ <u>where</u> $z \in \Lambda - \{a_1, a_2, a_3\}$ <u>span a dense subspace of</u> $A_2(\Omega)$.

<u>Remarks</u> 1. Obviously the theorem implies that the functions $\Phi^z(\zeta)$ are dense in $A_2(\Sigma)$ where Σ is any union of components of Ω, because in this case $A_2(\Sigma) \subset A_2(\Omega)$.

2. For generalizations of this theorem see Bers [2] or Kra [6].

<u>Proof:</u> Suppose ℓ is bounded linear functional on $A_2(\Omega)$. To prove the theorem, we must show that, if $\ell(\Phi^z) = 0$ for each $z \in \Lambda - \{a_1, a_2, a_3\}$, then $\ell(\varphi) = 0$ for all $\varphi \in A_2(\Omega)$. Of course, one can find a bounded measurable function μ with support Ω such that

$$(3.1) \qquad \ell(\varphi) = \iint_{\Omega} \varphi(\zeta)\mu(\zeta) d\zeta \wedge \overline{d\zeta} \quad \text{for all } \varphi \in A_2(\Omega).$$

Now we let F be a potential for the function μ. In particular, let

$$(3.2) \qquad F(z) = \iint_{\Omega} \Phi^z(\zeta)\mu(\zeta) d\zeta \wedge \overline{d\zeta}.$$

The hypothesis tells us that $F(z) = \ell(\Phi^z) = 0$ for all $z \in \Lambda - \{a_1, a_2, a_3\}$. Recall that by lemma 4, F is continuous and $\frac{\partial F}{\partial \bar{z}} = \mu$ in the sense of distributions. To show that $\ell(\varphi) = 0$, we try to argue as follows:

$$(3.3) \qquad \ell(\varphi) = \iint_{\Omega} \frac{\partial F}{\partial \bar{\zeta}} \varphi \, d\zeta \wedge \overline{d\zeta} = \iint_{\Omega} \frac{\partial (F\varphi)}{\partial \bar{z}} dz \wedge \overline{dz}$$

$$= - \iint_{\Omega} \bar{\partial}(F\varphi \, dz) = \int_{\partial\Omega} F\varphi \, dz = 0.$$

If we knew φ to be continuous and $\partial\Omega$ to be sufficiently smooth, the fact that $F = 0$ on $\partial\Omega$ would make this argument valid. However, these hypotheses are not satisfied.

To get around these difficulties, our first step is to translate all the variables in the problem by the action of a Möbius transformation B. We choose B so that $B(0) = a_1$, $B(1) = a_2$, and $B(\infty) = a_3$. By letting $\hat{\Gamma} = B^{-1}\Gamma B$ and $\hat{\Omega} = B^{-1}\Omega$, the theorem reduces to the statement that the functions

$$\Phi^z(\zeta) = \frac{1}{2\pi i} \frac{z(z-1)}{(\zeta - z)(\zeta)(\zeta - 1)}$$

span a dense subspace of $A_2(\Omega)$ where $0, 1, \infty \in \Lambda$ and $z \in \Lambda - \{0, 1, \infty\}$.

The function $F(z)$ is now given by

$$F(z) = \iint_\Omega \Phi^z(\zeta)\mu(\zeta)d\zeta \wedge \overline{d\zeta} \tag{3.4}$$

$$= \frac{z(z-1)}{2\pi i} \iint_\Omega \frac{\mu(\zeta)d\zeta \wedge \overline{d\zeta}}{(\zeta - z)(\zeta)(\zeta - 1)} .$$

<u>Lemma 6</u>. $F(z)$ defined by (3.4) has the following properties:

i) F is continuous on C,

ii) $\frac{\partial F}{\partial \bar{z}} = \mu$,

iii) $F(z) = 0(|z| \log|z|)$, $z \to \infty$, and

iv) for every $R > 0$, there exist $C(R)$ such that
$|F(z) - F(w)| \leq C(R)|z-w| \, \big|\log|z-w|\big|$ for all $|z|, |w| \leq R$.

This lemma is proved by the same methods used to prove Lemma 4.

Let $\delta(z)$ be the smaller of the number e^{-2} and the distance from z to $\partial\Omega$. Since $F(w) = 0$ for w in $\partial\Omega$, part (iv) of the above lemma tells us that $|F(z)| \leq C(R)\,\delta(z)\,|\log \delta(z)|$ for $|z| \leq R$. Note that this inequality tells us not only that F vanishes on $\partial\Omega$ but also puts a bound on the size of F for points near $\partial\Omega$. It is this fact which makes it possible to patch up the argument in (3.3) by the use of an ingenious device known as Ahlfors' mollifier.

Let $j(t)$ be a smooth function with $0 \leq j(t) \leq 1$ for all $t \in \mathbb{R}$ and $j(t) = 0$ for $t \leq 1$ and $j(t) = 1$ for $t \geq 2$. For each integer n and $z \in \Omega$, set $\omega_n(z) = j\left(\dfrac{n}{\log\log \frac{1}{\delta(z)}}\right)$. (Note that we arranged for $\delta(z) \leq e^{-2}$, so that the log log term in ω_n would be defined.) Moreover, since $|\delta(z) - \delta(w)| \leq |z-w|$, δ has generalized derivatives and we can compute $\dfrac{\partial\omega_n}{\partial\bar{z}}$ in the ordinary way.

$$\frac{\partial\omega_n}{\partial\bar{z}} = j'(\;''\;)\,\frac{n}{(\log\log\delta^{-1})^2}\cdot\frac{1}{\log\delta^{-1}}\cdot\frac{1}{\delta(z)}\cdot\frac{\partial\delta}{\partial\bar{z}}$$

Since $j'(t)$ and $\dfrac{\partial\delta}{\partial\bar{z}}$ are bounded, we conclude

$$\left|\frac{\partial\omega_n}{\partial\bar{z}}\right| \leq (\text{const.})\,\frac{n}{(\log\log\delta^{-1})^2}\cdot\frac{1}{\log\delta^{-1}}\cdot\frac{1}{\delta(z)}$$

Also we may assume $1 < n(\log\log\delta^{-1})^{-1} < 2$ since otherwise j' vanishes. Therefore we arrive at the inequality

$$\left|\frac{\partial\omega_n}{\partial\bar{z}}\right| \leq (\text{const.})\,\frac{1}{n}\,\frac{1}{\delta\,\log\delta^{-1}}. \tag{3.5}$$

It follows from the definition of ω_n that $\omega_n(z) = 0$ whenever

$\delta(z) \leq \exp(-e^n)$. Let $D(R) = \Omega \cap \{z \mid |z| < R\}$ and $\Gamma(R) = \Omega \cap \{z \mid |z| = R\}$. Applying Stoke's theorem, we find

$$(3.6) \qquad \iint_{D(R)} \omega_n \varphi\mu \, dz \wedge \overline{dz} = \iint_{D(R)} \omega_n \frac{\partial}{\partial \overline{z}}(\varphi F) dz \wedge \overline{dz}$$

$$= - \int_{\Gamma(R)} \omega_n \varphi F \, dz - \iint_{D(R)} \varphi F \frac{\partial \omega_n}{\partial \overline{z}} dz \wedge \overline{dz}.$$

By (3.5) and the fact that $|F(z)| \leq C(R)\delta(z)|\log \delta(z)|$, we see that

$$\left| \iint_{D(R)} \varphi F \frac{\partial \omega_n}{\partial \overline{z}} dz \wedge dz \right| \leq (\text{const.}) \iint_{D(R)} \left| \varphi(z) \frac{dz \wedge \overline{dz}}{n} \right|.$$

Since ω_n approaches 1 in bounded pointwise convergence, if we take the limit in (3.6) as $n \to \infty$, we find that

$$(3.7) \qquad \left| \iint_{D(R)} \varphi\mu \, dz \wedge \overline{dz} \right| \leq \left| \int_{\Gamma(R)} \varphi(z) F(z) dz \right|.$$

By part (iii) of lemma 6, the right hand integral is less than a constant times

$$(3.8) \qquad R \log R \int_{\Gamma(R)} |\varphi| \, |dz|.$$

Notice that $\int_0^\infty \{ \int_{\Gamma(R)} |\varphi| \, |dz| \} dR = \iint_{\Omega} |\varphi| \frac{|dz \wedge \overline{dz}|}{2} < \infty$.

If (3.8) is always larger than a positive number ϵ, then $\iint_{\Omega} |\varphi \, dz \wedge \overline{dz}| \geq \int_2^\infty \frac{\epsilon}{R \log R} dR = \infty$. Therefore, we conclude that

$$\liminf_{R \to \infty} \int_{\Gamma(R)} |F\varphi| \, |dz| = 0$$

from which it follows that $\iint_{\Omega} \varphi\mu \, dz \wedge \overline{dz} = 0$ whenever $\varphi \in A_2(\Omega)$.

This completes the proof.

Remark. It is unknown whether the analogous theorem to theorem 7 is valid for $q > 2$. See Kra [6].

§4. AHLFORS' FINITENESS THEOREM

Theorem 8. (Ahlfors [1]) If Γ is a finitely generated nonelementary Kleinian group, then $\Omega(\Gamma)/\Gamma$ is a finite union of Riemann surfaces of finite type.

Remark. This theorem does not say that $\Omega(\Gamma)$ has finitely many components. In most cases the number of components of $\Omega(\Gamma)$ is infinite.

The proof of the finiteness theorem depends on the following classical lemma.

Lemma 7. Let Δ be a component of $\Omega(\Gamma)$ and let Γ_1 be the subgroup of Γ which leaves Δ invariant. Then $A_2(\Delta, \Gamma_1)$ has finite dimension if and only if Δ/Γ_1 is of finite type.

Proof: By classical function theory (see Kra [6], page 324-328) one can construct an infinite dimensional space of square integrable abelian differentials on $S = \Delta/\Gamma_1$ if S has infinite genus or if S has infinitely many elliptic points or punctures. The product of any two of these abelian differentials will be in $A_2(\Delta, \Gamma_1)$. Conversely, if Δ/Γ_1 is of finite type then the classical Riemann-Roch theorem asserts that

$\dim A_2(\Delta, \Gamma_1) < \infty$ and gives an explicit formula for this dimension in terms of the signature of S (cf. Theorem 3).

Remark. It is possible for $\dim A_2(\Delta, \Gamma_1) = 0$. For example, when Δ/Γ_1 is a thrice punctured sphere with signature (2, 3, 7).

Corollary. If Γ is a finitely generated, Kleinian group, then each component of $\Omega(\Gamma)/\Gamma$ is a Riemann surface of finite type.

Proof: Let Σ be the union of all components of Ω which cover a single component of Ω/Γ. Let Δ be a single component of Σ and Γ_1 the subgroup Γ which leaves Δ invariant. Then obviously $B_2(\Delta, \Gamma_1) = B_2(\Sigma, \Gamma)$. By theorems 6 and 7, $\beta \circ i : B_2(\Sigma, \Gamma) \longrightarrow H^1(\Gamma, \prod_2)$ is injective. Since by lemma 3, $H^1(\Gamma, \prod_2)$ has finite dimension, we conclude $B_2(\Delta, \Gamma_1) = A_2(\Delta, \Gamma_1)$ and these spaces have finite dimension. Therefore Δ/Γ_1 (which equals Σ/Γ), is of finite type, by lemma 7.

To finish the proof of theorem 8, we must show that $\Omega(\Gamma)/\Gamma$ has finitely many components. To accomplish this, we prove the following theorem.

Theorem 9. *Let Γ be a nonelementary Kleinian group such that each component of Ω/Γ is of finite type. Then $\beta \circ i : B_q(\Omega, \Gamma) \longrightarrow H^1(\Gamma, \prod_{2q-2})$ is an injection.*

Proof: Let ω be a fundamental region in Ω for Γ. $\omega = \bigcup_{i=1}^{R} \omega_i$ where R is a positive integer or $R = \infty$ and each $\pi(\omega_i)$ yields precisely one of the components of Ω/Γ. Suppose $\psi \in B_q(\Omega, \Gamma)$ and $\beta \circ i(\psi) = 0$. By lemma 4, we know that the generalized Beltrami coefficient $\lambda^{2-2q}\overline{\psi}$ has a potential F such that $\gamma_{1-q}^{*} F = F$ for each $\gamma \in \Gamma$. Let $\psi_i(z) = \psi(z)$ if $z \in \Gamma\omega_i$ and $\psi_i(z) = 0$ otherwise. We must show that $\psi_i = 0$ for each i.

$$\iint_{\omega_i} |\psi_i|^2 \lambda^{2-2q} |dz \wedge dz| = \iint_{\omega_i} \psi_i \frac{\partial F}{\partial \overline{z}} |dz \wedge \overline{dz}|$$

$$= \iint_{\omega_i} \frac{\partial}{\partial \overline{z}} (F\psi_i) |dz \wedge \overline{dz}|.$$

Notice that $F\psi_i\, dz$ is invariant. If there are no parabolic punctures on $\pi(\omega_i)$, one can apply Stokes' theorem and this last integral becomes

$$-\frac{1}{2} \int_{\partial\omega_i} F\psi_i \, dz = 0. \tag{4.1}$$

If $\pi(\omega_i)$ has a parabolic puncture, then by lemma 1, one can assume there is a parabolic element $\gamma \in \Gamma$ of the form $\gamma(z) = z + 1$ and a constant c such that $\Gamma\omega_i \supseteq \{z \mid 0 \leq \operatorname{Re} z < 1 \text{ and } \operatorname{Im} z > c\}$. One can use a similar Stokes' theorem argument to show that $\int_{\omega_i} F\psi_i\, dz = 0$ if one shows that

$$\lim_{b \to \infty} \int_0^1 \psi(x + ib) F(x + ib) dx = 0. \tag{4.2}$$

But by (2.2) we know $F(x + ib) = 0((x^2 + b^2)^{q-1})$ and because φ is a cusp form we know that $|\varphi(x + ib)| \leq (\text{const.}) e^{-2\pi b}$. Hence the limit in (4.2) is zero.

Corollary 1. $\dim B_q(\Omega, \Gamma) \leq \dim H^1(\Gamma, \prod_{2q-2})$.

Corollary 2. If Γ is generated by N elements

$$\sum_{j=1}^{R} \{(2q-1)(g_j-1) + \sum_{p \in \hat{S}_j} [q - \frac{q}{\nu(p)}]\} \leq (2q-1)(N-1).$$

Here the sum is taken over all elliptic or parabolic points p of $\Omega(\Gamma)/\Gamma$ and g_j is the genus of the component S_j of Ω/Γ.

Corollary 2 follows from corollary 1, lemma 3 and theorem 3.

Corollary 3. If Γ is generated by N elements and the number of components of $\Omega(\Gamma)/\Gamma$ is R, then $R \leq 18(N-1)$.

Proof: By elementary methods one shows that $\dim A_4(\Delta_j, \Gamma) + \dim A_6(\Delta_j, \Gamma) \geq 1$ for all j. Adding the inequality in corollary 2 for $q = 4$ and $q = 6$, the result follows. This corollary completes the proof of Ahlfors' finiteness theorem.

§5. THE AREA THEOREMS

Theorem 10. (Bers' 1st area theorem [4]) *Area* $(\Omega/\Gamma) \leq 4\pi(N-1)$ *where the area is Poincaré area and* N = *the number of generators of* Γ.

Proof: Merely multiply the inequality in corollary 2 to theorem 9 by $\frac{2\pi}{q}$ and let $q \to \infty$. The left hand side approaches the Poincaré area and the right hand side approaches $4\pi(N-1)$.

<u>Theorem 11</u>. (Bers' 2nd area theorem [4]) <u>Suppose</u> Ω_1 <u>and</u> Ω_2 <u>are nonempty open subsets of</u> Ω, <u>each of which is invariant under</u> Γ, <u>and suppose</u> $\Omega_1 \cup \Omega_2 = \Omega$ <u>and</u> $\Omega_1 \cap \Omega_2 = \emptyset$. <u>If</u> Ω_1 <u>is connected, then</u> $\underline{\text{area}}(\Omega_2/\Gamma) \leq \underline{\text{area}}(\Omega_1/\Gamma)$.

<u>Proof:</u> Our method will be to establish the existence of an injective mapping $L : B_q(\Omega_2, \Gamma) \longrightarrow B_q(\Omega_1, \Gamma)$. Just as in the proof of the first area theorem, the inequality for areas will follow multiplying the corresponding inequality for dimensions of cohomology groups by $\frac{2\pi}{q}$ and letting $q \to \infty$.

To construct the mapping, let $\varphi \in B_q(\Omega_2, \Gamma)$ and let $\mu(z) = \lambda^{2-2q}(z)\overline{\varphi}(z)$ for $z \in \Omega_2$ and $\mu(z) = 0$ for $z \in C - \Omega_2$. Let F be a potential for the generalized Beltrami coefficient μ and define $L(\varphi)(z) = \frac{d^{2q-1}}{dz^{2q-1}} F(z)$ for $z \in \Omega_1$. It is clear that $L(\varphi)$ will be a holomorphic function in Ω_2, that $L(\varphi)$ depends only on φ and not on which potential F is chosen, and that L is an antilinear mapping. By differentiating the integral expression for F $(2q-1)$ times, one finds that

$$L(\varphi)(z) = \frac{(2q-1)!}{2\pi i} \iint_{\Omega_1} \frac{\lambda^{2-2q}(\zeta)\overline{\varphi}(\zeta)}{(\zeta - z)^{2q}}\, d\zeta \wedge \overline{d\zeta}.$$

Using this fact and the formula $(A\zeta - Az)^2 = (\zeta - z)^2 A'(\zeta)A'(z)$, it is easy to show that $L(\varphi)$ is a differential form of weight $(-2q)$. Let $\psi = L(\varphi)$. To show that $\psi \in B_q(\Omega_2)$ we must show that if a puncture is realized by the parabolic transformation $\gamma(z) = z + 1$ with

$\Omega_2 \supset \{z \,|\, \mathrm{Im}\, z > c\}$, then $\psi(z) = 0(e^{-2\pi y})$. But $\psi(z) = \sum_{-\infty}^{\infty} a_n e^{2\pi i n z}$ for $\mathrm{Im}\, z > c$. By the fact that $\psi(z) = \frac{d^{2q-1}}{dz^{2q-1}} F(z)$ one finds that $F(z) = \sum_{-\infty}^{\infty} a_n (2\pi i n)^{1-2q} e^{2\pi i n z} + a_0 z^{2q-1} + v$ for $\mathrm{Im}\, z > 0$ where $v \in \prod_{2q-2}$. But $F(z) = 0(|z|^{2q-2})$. Therefore, $a_n = 0$ for $n \leq 0$ and hence $\psi = \sum_{n=1}^{\infty} a_n e^{2\pi i n z} = 0(e^{-2\pi y})$ as $y \to \infty$.

<u>Theorem 12.</u> <u>If</u> Ω_1 <u>is connected, then</u> $L : B_q(\Omega_2, \Gamma) \longrightarrow B_q(\Omega_1, \Gamma)$ <u>is injective.</u>

<u>Proof:</u> Suppose $L(\varphi) = 0$. Then $\lambda^{2-2q}\bar{\varphi}$ has a potential F_0 such that F_0 is a polynomial of degree $\leq 2q-2$ in Ω_1. Let v_0 be this polynomial and let $F = F_0 - v_0$. Hence F is a potential which vanishes on Ω_1. Since Ω_1 is connected $\Lambda = \partial\Omega_1$. Thus, we have, by continuity, that $F|\Lambda = 0$. Hence, by lemma 5, $\beta \circ i(\varphi) = 0$. By theorem 9, $\varphi = 0$.

REFERENCES

[1] L.V. Ahlfors, Finitely generated Kleinian groups, <u>Amer. J. Math.</u>, <u>86</u> (1964), 413-429 and <u>87</u> (1965), 759.

[2] L. Bers, An approximation theorem, <u>J. Analyse Math.</u>, <u>14</u> (1965), 1-4.

[3] L. Bers, On Ahlfors' finiteness theorem, <u>Amer. J. Math.</u>, <u>89</u> (1967), 113-134.

[4] L. Bers, Inequalities for finitely generated Kleinian groups, <u>J. Analyse Math.</u>, <u>18</u> (1967), 23-41.

[5] L. Bers, Automorphic forms and Poincaré series for infinitely generated Fuchsian groups, _Amer. J. Math._, _87_ (1965), 196-214.

[6] I. Kra, Automorphic Forms and Kleinian Groups, W.A. Benjamin (1972).

4. DEFORMATION SPACES*

Irwin Kra
SUNY at Stony Brook

Let G be a (non-elementary) finitely generated Kleinian group. In this chapter we show how the set $\hat{T}(G)$ of all "marked" Kleinian groups quasiconformally equivalent to G forms a finite dimensional complex analytic manifold, and the set $\hat{R}(G)$ of conjugacy classes of Kleinian groups quasiconformally equivalent to G forms a normal complex space. This deformation theory contains as a special case and is completely dependent on Teichmüller space theory for Riemann surfaces (of finite type) or equivalently on Teichmüller space theory for finitely generated Fuchsian groups of the first kind.

We fix once and for all a finitely generated (non-elementary) Kleinian group G, and let Σ be an invariant union of components for G. The region of discontinuity of G will be denoted by Ω. All Kleinian groups under consideration will be assumed finitely generated. Almost all Fuchsian groups considered will be finitely generated, of the first kind.

§1. QUASICONFORMAL DEFORMATIONS OF KLEINIAN GROUPS

Let f be a quasiconformal automorphism of the complex

*Work partially supported by NSF grant GP-19572.

sphere $\hat{\mathbb{C}}$. We say that f is <u>compatible</u> with G provided the Beltrami coefficient μ of f vanishes on the limit set $\Lambda = \Lambda(G)$ of G, and fGf^{-1} is again a Kleinian group. This condition is equivalent to μ being a <u>Beltrami coefficient for</u> G; that is,

i) $\mu(gz)\overline{g'(z)}/g'(z) = \mu(z)$, all $g \in G$, a.e. $z \in \hat{\mathbb{C}}$,

ii) $\mu|\Lambda = 0$, and

iii) $\|\mu\| = \text{ess sup } |\mu| < 1$.

WARNING. All quasiconformal mappings (compatible with G) to be considered will be assumed to be conformal on the limit set (of G). It is not known whether this is automatically satisfied by all finitely generated groups. (See Bers' lecture.)

We denote (as in Earle's lecture) by $M(G,\Sigma)$ the set (open unit ball in a closed linear subspace of $L_\infty(\mathbb{C})$) of Beltrami coefficients for G with support in Σ. Since $M(G,\Sigma)$ is an open subset of a complex Banach space, it has a natural topology and complex structure.

To every $\mu \in M(G,\Sigma)$ there corresponds (Ahlfors-Bers [4]) a unique normalized (fixing $0,1,\infty$) μ-conformal automorphism of $\hat{\mathbb{C}}$ denoted by w^μ.

DEFINITION AND LEMMA 1.1. <u>A Beltrami coefficient</u> $\mu \in M(G,\Sigma)$ <u>is called trivial if it satisfies one (hence both) of the following conditions</u>:

a) $w^\mu \circ g \circ (w^\mu)^{-1} = g$, <u>all</u> $g \in G$,

b) $w^\mu(z) = z$, *all* $z \in \Lambda$, *the limit set of* G.

A trivial Beltrami coefficient fixes each component of G as a consequence of the following elementary topological

LEMMA 1.2. *An orientation preserving automorphism of* $\hat{\mathbb{C}}$ *maps every component of the complement of its fixed point set onto itself.*

The set of trivial Beltrami coefficients for G with support in Σ is denoted by $M_0(G,\Sigma)$. It acts as a group of holomorphic mappings known as *right translations* on $M(G,\Sigma)$ by

$$M(G,\Sigma) \times M_0(G,\Sigma) \ni (\nu,\mu) \mapsto \nu\mu^{-1} \in M(G,\Sigma)$$

where

$$w^{\nu\mu^{-1}} = w^\nu \circ (w^\mu)^{-1}.$$

The quasiconformal deformation space, of G *with support in* Σ, is

$$\hat{T}(G,\Sigma) = M(G,\Sigma)/M_0(G,\Sigma)$$

endowed with the quotient topology and quotient complex structure (of a ringed space).

An element $\mu \in M(G,\Sigma)$ is called *strongly trivial* provided w^μ is homotopic to the identity on each component Δ of Σ modulo the ideal boundary of Δ. It is easy to show that the set of strongly trivial Beltrami coefficients, $\tilde{M}_0(G,\Sigma)$, forms a normal subgroup of $M_0(G,\Sigma)$. The *strong deformation space, of* G *with support in* Σ, is defined as

$$\tilde{T}(G,\Sigma) = M(G,\Sigma)/\tilde{M}_o(G,\Sigma).$$

If we set

$$\mathfrak{J}(G,\Sigma) = M_o(G,\Sigma)/\tilde{M}_o(G,\Sigma),$$

then it is clear that

$$\hat{T}(G,\Sigma) \cong \tilde{T}(G,\Sigma)/\mathfrak{J}(G,\Sigma).$$

If $\Sigma = \Omega$, the entire region of discontinuity we write $\hat{T}(G) = \hat{T}(G,\Omega)$, $\tilde{T}(G) = \tilde{T}(G,\Omega)$, and $\mathfrak{J}(G) = \mathfrak{J}(G,\Omega)$.

§2. THE DECOMPOSITION THEOREM

Let $\Delta_1, \Delta_2, \ldots, \Delta_r$ be a complete list of non-conjugate components of Σ and $G_j = G_{\Delta_j}$, the stability subgroup of Δ_j in G. By Ahlfors' Finiteness Theorem [3] (see Gardiner's lecture) this is indeed a finite list and Δ_j/G_j is a surface of finite type. Furthermore, each G_j is finitely generated. The study of deformation spaces of Kleinian groups reduces to the study of deformation spaces of <u>function</u> groups (Kleinian groups with an invariant component) as a result of the following

THEOREM 2.1 (Kra [22]). <u>We have</u>

$$\tilde{T}(G,\Sigma) \cong \tilde{T}(G_1,\Delta_1) \times \ldots \times \tilde{T}(G_r,\Delta_r),$$

<u>and</u>

$$\hat{T}(G,\Sigma) \cong \hat{T}(G_1,\Delta_1) \times \ldots \times \hat{T}(G_r,\Delta_r).$$

The first isomorphism is trivial. The second is a consequence of Lemma 1.2 and another elementary

LEMMA 2.2. *Let D be an open subset of* $\hat{\mathbb{C}}$ *and f a topological homeomorphism of* $\hat{\mathbb{C}}$ *that is quasiconformal on D and such that* $f|\hat{\mathbb{C}}\setminus D = \mathrm{id}$. *Then f is quasiconformal and its Beltrami coefficient is supported in D.*

Lemma 2.2 together with some arguments specific to Kleinian groups leads to the following extension

THEOREM 2.3 (Maskit [25]). *Let f be a topological automorphism of* Σ *such that* $f \circ g = g \circ f$, *all* $g \in G$. *Then f is the restriction of a global topological automorphism F of* $\hat{\mathbb{C}}$ *with* $F|\hat{\mathbb{C}}\setminus\Sigma = \mathrm{id}$ *and* $F \circ g = g \circ F$, *all* $g \in G$. *Further, F is quasiconformal (conformal) if f is.*

§3. REDUCTION TO THE FUCHSIAN CASE

Let Δ be a component of Σ. Let $h : U \to \Delta$ be a holomorphic universal covering map. Let Γ be the *Fuchsian model of G over* Δ; that is, the group of conformal automorphisms γ of U such that there is a $\chi(\gamma) \in G_\Delta$ with $h \circ \gamma = \chi(\gamma) \circ h$. We then have an exact sequence of Kleinian groups

$$\{1\} \longrightarrow H \xrightarrow{\text{inj}} \Gamma \xrightarrow{\chi} G_\Delta \longrightarrow \{1\}$$

with H (the covering group) and Γ Fuchsian, and Γ and G_Δ of finite type (as a matter of fact, $U/\Gamma \cong \Delta/G_\Delta$ as Riemann surfaces with ramification points).

REMARK. The Fuchsian model of G over Δ used above differs from the one defined by Maskit [25]. However, both models lead to equivalent results (Theorem 7.1).

For $\mu \in M(\Gamma) = M(\Gamma,U)$ we define $h^*\mu \in M(G_\Delta,\Delta)$ by $(h^*\mu)\circ h = \mu h'/\overline{h'}$, and verify almost trivially that

$$(3.1) \qquad h^* : M(\Gamma) \to M(G_\Delta,\Delta)$$

is a (linear, whenever possible) norm preserving (surjective) isomorphism. Furthermore, we have the following

PROPOSITION 3.1. *The map* h^* *of* (3.1) *induces holomorphic surjections*

$$T(\Gamma) = \hat{T}(\Gamma,U) \to \hat{T}(G_\Delta,\Delta)$$

$$T(\Gamma) \overset{\cong}{\to} \tilde{T}(G_\Delta,\Delta)$$

(*with the second map being an isomorphism*).

REMARK. The space $T(\Gamma)$ is called the *Teichmüller space* for the Fuchsian group Γ.

Before we can continue the investigation of the above mappings, we must turn to a more careful examination of

§4. THE FUCHSIAN CASE

For Γ Fuchsian, we denote by $B(\Gamma) = B_2(\Gamma,L)$ the Banach space of bounded automorphie forms of weight -4 for Γ with support in the lower half plane, L (see Gardiner's lecture). For $\mu \in M(\Gamma)$, let φ^μ be the Schwarzian derivative of

$f = w^\mu | L$ (the definition makes sense since w^μ is holomorphic and univalent on L),

$$\varphi^\mu = (\frac{f''}{f'})' - \frac{1}{2}(\frac{f''}{f'})^2 .$$

A result of Nehari [26] shows that $\varphi^\mu \in B(\Gamma)$. Lemma 1.1 shows that φ^μ depends only on the equivalence class of μ in $T(\Gamma)$. Furthermore, the induced mapping (from $T(\Gamma)$ into $B(\Gamma)$) is easily seen to be injective. We may view $\mu \mapsto \varphi^\mu$ as a holomorphic mapping

$$\Phi : M(\Gamma) \to B(\Gamma).$$

The derivative of this mapping (at the origin) can be computed and seen (Bers [8]) to be surjective. Since $B(\Gamma)$ is a finite dimensional space, it can be shown using the implicit function theorem that $\Phi(M(\Gamma)) = \Phi(T(\Gamma))$ is open in $B(\Gamma)$. We have the following

THEOREM 4.1 (Bers [8]). *The Teichmüller space* $T(\Gamma)$ *has a unique complex structure so that*

$$\Phi : M(\Gamma) \to T(\Gamma)$$

is holomorphic with local holomorphic sections. We can realize $T(\Gamma)$ *as a bounded domain (of holomorphy) in* $B(\Gamma)$. *Further* $T(\Gamma)$ *is topologically a cell.*

The fact that $T(\Gamma)$ is a domain of holomorphy was proven by Bers-Ehrenpreiss [12]. (See also Royden's lecture.) Teichmüller's theorem (see for example Ahlfors [2] or

Bers [6]) implies that every $\mu \in M(\Gamma)$ is equivalent to a unique Teichmüller coefficient; that is, a Beltrami coefficient μ with

$$\mu(z) = k\,\varphi(\bar{z})/|\varphi(\bar{z})|,\ z \in U,$$

$k \in \mathbb{R}$, $0 \leq k < 1$, $0 \neq \varphi \in B(\Gamma)$.

It follows easily from this result that $T(\Gamma) \cong B(\Gamma)$ (as topological manifolds).

Bers-Greenberg [13] show that $T(\Gamma)$, as a complex manifold, depends only on the type of Γ. (This result has been obtained independently by Marden [23]. For a rather simple proof see Earle-Kra [17].) We thus define $T(g,n)$ as the Teichmüller space $T(\Gamma)$ for some group Γ of type (g,n).

§5. MODULAR GROUPS

An automorphism θ of the Kleinian group G is called geometric, with respect to Σ, if and only if there is a quasiconformal automorphism f of $\hat{\mathbb{C}}$ compatible with G such that

$$f\Sigma = \Sigma$$

and

$$\theta(\gamma) = f \circ \gamma \circ f^{-1},\ \text{all } \gamma \in G.$$

The map f induces a biholomorphic automorphism of $M(G,\Sigma)$ by sending $\mu \in M(G,\Sigma)$ into the Beltrami coefficient of $w^{\mu} \circ f^{-1}|\Sigma$. It is easy to check that this mapping preserves equivalence classes and hence induces a biholomorphic self-mapping

$$\theta^* : \hat{T}(G,\Sigma) \to \hat{T}(G,\Sigma)$$

which depends only on the automorphism θ, in fact only on the conjugacy class of θ modulo inner automorphisms of G. We thus define the modular group Mod(G,Σ) as the quotient of the group of geometric (with respect to Σ) automorphisms by the subgroup of inner automorphisms. (The action of Mod(G,Σ) on $\hat{T}(G,\Sigma)$ is not always effective.)

REMARK. The above definition of Mod(G,Σ) differs from the definition used by Bers [11].

The moduli space is defined by

$$\hat{R}(G,\Sigma) = \hat{T}(G,\Sigma)/\text{Mod}(G,\Sigma).$$

As usual we set

$$\hat{R}(G) = \hat{R}(G,\Omega),$$

and observe that it denotes the set of conjugacy classes of Kleinian groups quasiconformally equivalent to G.

If Γ is a Fuchsian group (operating on U), we set Mod Γ = Mod(Γ,U). Further, Mod Γ depends only on the signature of Γ. We define Mod(g,n) to be Mod Γ for a group of signature $(g,n;\infty,\ldots,\infty)$. If Γ has type (g,n) then Mod Γ is a subgroup of finite index in Mod(g,n).

The group Mod(g,n) acts discontinuously on $T(g,n)$ (a classical result, for the proof see, for instance, Bers [10]) and thus the Riemann space (for Γ)

$$R(\Gamma) = \hat{R}(\Gamma,U)$$

is a normal complex space (as a result of a theorem of Cartan

[14]). For more about the group Mod(g,n) see Royden's lecture.

REMARK. If Γ is Fuchsian, then $R(\Gamma)$ consists of conjugacy classes of Fuchsian groups with the same signature as Γ. This assertion follows from the following:

1) Two Fuchsian groups are quasiconformally equivalent if and only if they have the same signature.
2) For each $\mu \in M(\Gamma)$ there exists a unique normalized μ-conformal automorphism of U denoted by w_μ (Ahlfors-Bers [4]). This automorphism conjugates Γ onto another Fuchsian group.
3) For $\mu,\nu \in M(\Gamma)$, μ and ν are equivalent (in $T(\Gamma)$) if and only if $w_\mu|\mathbb{R} = w_\nu|\mathbb{R}$.
4) The modular group Mod Γ can be defined by looking at quasiconformal automorphisms f of U (all such f have global compatible extensions) that conjugate Γ onto itself.

Note also that T(g,n) represents "marked" Riemann surfaces of type (g,n), while R(g,n) = T(g,n)/Mod(g,n) represents the conformal equivalence classes of surfaces of type (g,n).

§6. THE FUNCTION GROUP CASE

We return to the situation treated in §3. If ω is a quasiconformal automorphism of U with

(6.1) $\omega H \omega^{-1} = H$, $\omega \Gamma \omega^{-1} = \Gamma$,

then ω induces a quasiconformal automorphism f of Δ such that

$$f \circ h = h \circ \omega$$

and

(6.2) $f G_\Delta f^{-1} = G_\Delta$.

Further, every quasiconformal automorphism f of Δ that satisfies (6.2) is so induced (ω is unique except that it may be replaced by $\gamma \circ \omega$, $\gamma \in H$).

If f and ω are as above, then

(6.3) $f \circ g \circ f^{-1} = g$, all $g \in G_\Delta$

if and only if

(6.4) $\omega \circ \gamma \circ \omega^{-1} \circ \gamma^{-1} \in H$, all $\gamma \in \Gamma$.

It follows from Theorem 2.3 (and Lemma 2.2) that such an f is the restriction to Δ of a global compatible f (that satisfies (6.2)) and whose Beltrami coefficient is supported on $G\Delta = \{g(\Delta); g \in G\}$.

Let $\mathrm{Mod}^H(\Gamma)$ denote the set of elements in Mod Γ induced by all ω satisfying (6.1), and let $\mathrm{Mod}_H(\Gamma)$ the subset of induced by all ω satisfying (6.4). Both of these are groups. Further, $\mathrm{Mod}_H(\Gamma)$ is a normal subgroup of $\mathrm{Mod}^H(\Gamma)$.

PROPOSITION 6.1. _The group_ $\mathrm{Mod}_H(\Gamma)$ _acts freely on_ $T(\Gamma)$. _Furthermore_,

$$T(\Gamma)/\mathrm{Mod}_H(\Gamma) \cong \hat{T}(G_\Delta, \Delta),$$

<u>and thus</u> $\hat{T}(G_\Delta,\Delta)$ <u>is a finite dimensional complex analytic manifold with holomorphic universal covering space</u> $T(\Gamma)$.

§7. THE GENERAL CASE

We now consider the situation described in §2. Let $\Gamma_j(j=1,\ldots,r)$ be the Fuchsian model of G over Δ_j defined by the holomorphic universal covering map $h_j : U \to \Delta_j$ with covering group $H_j \subset \Gamma_j$. We define

$$\mathrm{Mod}^*(G,\Sigma) = \mathrm{Mod}^{H_1}\Gamma_1 \times\ldots\times \mathrm{Mod}^{H_r}\Gamma_r$$

$$\mathrm{Mod}_*(G,\Sigma) = \mathrm{Mod}_{H_1}\Gamma_1 \times\ldots\times \mathrm{Mod}_{H_r}\Gamma_r.$$

Proposition 6.1 implies almost immediately the following

THEOREM 7.1 (Maskit [25]). <u>We have</u>

$$\tilde{T}(G,\Sigma) \cong T(\Gamma_1) \times\ldots\times T(\Gamma_r),$$

<u>and is hence simply connected. Further,</u>

$$\hat{T}(G,\Sigma) \cong (T(\Gamma_1) \times\ldots\times T(\Gamma_r))/\mathrm{Mod}_*(G,\Sigma).$$

COROLLARY 1. <u>The deformation space</u> $\hat{T}(G,\Sigma)$ <u>is a complex analytic manifold with universal covering space</u> $\tilde{T}(G,\Sigma)$.

COROLLARY 2. <u>If each component of</u> Σ <u>is simply connected, then</u>

$$\tilde{T}(G,\Sigma) \cong \hat{T}(G,\Sigma) \cong T(\Gamma_1) \times\ldots\times T(\Gamma_r).$$

We can also obtain

THEOREM 7.2 (Bers [11]). <u>The group</u> $\mathrm{Mod}(G,\Sigma)$ <u>is a group of holomorphic automorphisms of</u> $\hat{T}(G,\Sigma)$. <u>It acts properly</u>

discontinuously on $\hat{T}(G,\Sigma)$, *and* $\hat{R}(G,\Sigma)$ *is thus a normal complex space.*

The group $\mathrm{Mod}(G,\Sigma)$ is induced by quasiconformal automorphisms f of $\hat{\mathbb{C}}$ that conjugate G into itself and fix Σ. There is thus a normal subgroup $\mathrm{Mod}_o(G,\Sigma)$ of finite index in $\mathrm{Mod}\,(G,\Sigma)$ that is induced by quasiconformal automorphisms of $\hat{\mathbb{C}}$ that fix each $\Sigma_j = \{g(\Delta_j); g \in G\}$. Let f be such an automorphism of $\hat{\mathbb{C}}$. For each j, there is a $g_j \in G$ such that $f_j = g_j^{-1} \circ f$ fixes Δ_j and $f_j G_j f_j^{-1} = G_j$. By the introductory remarks of §6, there is an automorphism ω_j of U that satisfies

$$f_j \circ h_j = h_j \circ \omega_j$$

and

$$\omega_j H_j \omega_j^{-1} = H_j,\ \omega_j \Gamma_j \omega_j^{-1} = \Gamma_j.$$

Further, ω_j determines a well defined element of $\mathrm{Mod}^{H_i}(\Gamma_i)$. If f induces the identity automorphism of G, then the element of $\mathrm{Mod}^{H_i}(\Gamma_i)$ determined by the above procedure actually lands in $\mathrm{Mod}_{H_i}(\Gamma_i)$. We thus obtain

PROPOSITION 7.3. *There is a normal subgroup of finite index of* $\mathrm{Mod}(G,\Sigma)$ *which is isomorphic to a subgroup of* $\mathrm{Mod}^*(G,\Sigma)/\mathrm{Mod}_*(G,\Sigma)$.

Proposition 7.3 easily implies Theorem 7.2.

§8. THE COMPLEX STRUCTURE IS CANONICAL

We have shown that the space of marked Riemann surfaces of genus $g(\geq 2)$, $T(g,0)$, is a complex analytic manifold. In what sense is the complex structure canonical?

Let X be a compact Riemann surface of genus $g \geq 2$. Let $(a_1,\ldots,a_g,b_1,\ldots,b_g)$ be a canonical homology basis for X; that is, the intersection matrix

$$\begin{pmatrix} a_i \cdot a_j & a_i \cdot b_j \\ b_i \cdot a_j & b_i \cdot b_j \end{pmatrix} \qquad i,j = 1,\ldots,g$$

(where $a \cdot b$ denotes the intersection number of the curves a and b) is of the form

$$\begin{pmatrix} 0 & I \\ -I & 0 \end{pmatrix}$$

(where 0 is the $g \times g$ zero matrix, and I is the $g \times g$ identity matrix). Let $\omega_1,\ldots,\omega_g$ be a canonical basis for abelian differentials of the first kind dual to the given homology basis; that is,

$$\int_{a_i} \omega_j = \begin{cases} 0 \text{ if } i \neq j \\ 1 \text{ if } i = j. \end{cases}$$

Consider the $g \times g$ <u>period matrix</u> π

$$\pi = (\pi_{ij}),\ i,j = 1,\ldots,g,$$

where

$$\pi_{ij} = \int_{b_i} \omega_j .$$

A classical result (due to Riemann) asserts that π is a point in

$\mathcal{S}_g$, *the Siegel upper half space of genus* g; that is, the set of g×g (complex) symmetric matrices whose imaginary part is positive definite.

We choose a model $T(\Gamma)$ for $T(g,0)$ with Γ a Fuchsian group of type $(g,0)$. The curves $a_1,\ldots,a_g,b_1,\ldots,b_g$ determine elements $A_1,\ldots,A_g,B_1,\ldots,B_g$ in Γ (every homology basis lifts to a homotopy basis and $\pi_1(U/\Gamma) \cong \Gamma$). It is not hard to see that given any point $x \in T(\Gamma)$, with x the equivalence class of $\mu \in M(\Gamma)$, then the elements

$$A_1^\mu,\ldots,A_g^\mu,\ B_1^\mu,\ldots,B_g^\mu$$

in $\Gamma_\mu = w_\mu \Gamma w_\mu^{-1}$ (where $A_i^\mu = w_\mu \circ A_i \circ w_\mu^{-1}$, $B_j^\mu = w_\mu \circ B_j \circ w_\mu^{-1}$ $i,j = 1,\ldots,g$) project to a canonical homology basis on U/Γ_μ. In this way one obtains a canonical mapping (once a canonical homology basis has been chosen on U/Γ)

$$(8.1) \qquad S : T(g,0) \to \mathcal{S}_g .$$

Now $T(g,0)$ is a 3g-3 (= dim $B(\Gamma)$ by Riemann-Roch) dimensional manifold and $\mathcal{S}_g$ is a $(1/2)g(g+1)$ dimensional complex manifold and we have the following

THEOREM 8.1 (Rauch [27], Bers [7]). *The mapping* S *of* (8.1) *is holomorphic*. *It is of maximal rank* (*therefore, locally injective*) *except at a point* $x \in T(g,0)$ *representing a hyperelliptic Riemann surface*.

COROLLARY 1 (Rauch [27]). *In a neighborhood of a non-hyperelliptic Riemann surface*, 3g-3 *periods* (*that is, entries in* *the matrix π) can be used as local coordinates for* $T(g,0)$.

COROLLARY 2 (Bers [7]). *However, no fixed set of* $3g-3$ *periods can be used as local coordinates for all non-hyperelliptic surfaces.*

Thus, in a very real sense, the complex structure on $T(g,0)$ is natural. Of course, we have just touched on the very difficult classical problem of Schottky of determining the image $S(T(g,0))$ in $\mathfrak{s}_g$. For partial (but deep and important) results see the work of Andreotti-Mayer [2] and Farkas-Rauch [18].

Let G be a finitely generated Kleinian group. In what sense is the complex structure on $\hat{T}(G)$ canonical? The answer here is easier than in the classical case. Let $\gamma_1,\ldots,\gamma_r$ be a set of generators for G. Every homomorphism χ from G into the Möbius group, Möb, determines a point

$$(\chi(\gamma_1),\ldots,\chi(\gamma_r)) \in \text{Möb}^r.$$

Let Hom denote the set of parabolic homomorphisms of G into Möb (that is, those homomorphisms χ with $\operatorname{trace}^2\chi(\gamma) = 4$ for every parabolic element $\gamma \in G$ determined by a puncture on Ω/G). It is quite easy to see (Bers [9]) that Hom is an affine algebraic variety and that there is a holomorphic mapping

$$L : \hat{T}(G) \to \text{Hom}, \tag{8.2}$$

where $L(\Phi(\mu)) = (w^\mu\circ\gamma_1\circ(w^\mu)^{-1},\ldots,w^\mu\circ\gamma_r\circ(w^\mu)^{-1})$, $\mu \in M(G)$.

Is $L(\hat{T}(G))$ a submanifold of Hom? The question is completely open. Partial results have been obtained (Bers [9], Marden [24]). Is $L(\hat{T}(G))$ a manifold in a neighborhood of $L(0)$? Even here only partial results are known (Abikoff [1], Bers [9], Gardiner-Kra [19], Marden [24]).

The Fuchsian case is much simpler. Let Γ be a Fuchsian group and $Hom_{\mathbb{R}}$, the affine algebraic variety (over $\mathbb{R}$) of parabolic homomorphisms of Γ into the real Möbius group, $Möb_{\mathbb{R}}$. In analogy to (8.2), there is a real analytic diffeomorphism

$$L : T(\Gamma) \to Hom_{\mathbb{R}},$$

with $L(T(\Gamma))$ a submanifold of $Hom_{\mathbb{R}}$ (see, for instance, Gardiner-Kra [19]).

§9. A FUNCTORIAL APPROACH

Let Γ be a finitely generated Fuchsian group of the first kind operating on the upper half plane U. For each $\mu \in M(\Gamma)$, the domain $w^{\mu}(U)$ depends on $\Phi(\mu)$, the equivalence class of μ in $T(\Gamma)$. We may therefore form the <u>Bers fiber space</u>

$$F(\Gamma) = \{(\Phi(\mu),z) \in T(\Gamma) \times \mathbb{C};\ \mu \in M(\Gamma),\ z \in w^{\mu}(U)\},$$

and the <u>punctured fiber space</u>

$$F_0(\Gamma) = \{(\Phi(\mu),z) \in F(\Gamma);\ \mu \in M(\Gamma),\ z \in w^{\mu}(U_{\Gamma})\},$$

where U_{Γ} is the set of points in U which are not fixed by any elliptic element of Γ.

The group Γ acts discontinuously on $F(\Gamma)$ as a group of biholomorphic mappings by

$$\gamma(\Phi(\mu),z) = (\Phi(\mu),\ \gamma^{\mu}z)$$

where $\mu \in M(\Gamma)$, $z \in w^{\mu}(U)$, $\gamma \in \Gamma$, and

$$\gamma^{\mu} \circ w^{\mu} = w^{\mu} \circ \gamma .$$

The quotient space

$$V(\Gamma) = F(\Gamma)/\Gamma$$

is a complex manifold equipped with a projection π onto $T(\Gamma)$. The punctured fiber space $F_0(\Gamma)$ is invariant under Γ; in fact, $F_0(\Gamma)$ is the largest subset of $F(\Gamma)$ on which Γ acts freely. We form the punctured quotient space

$$V(\Gamma)' = F_0(\Gamma)/\Gamma .$$

For each point $\Phi(\mu) \in T(\Gamma)$, the fiber $\pi^{-1}(\Phi(\mu))$ in $V(\Gamma)'$ is the Riemann surface $w^{\mu}(U_{\Gamma})/w^{\mu}\Gamma(w^{\mu})^{-1}$. The complex manifold $V(\Gamma)'$ (the <u>n-punctured Teichmüller curve</u>) depends only on the type of Γ, so we define $V(g,n)'$ to be $V(\Gamma)'$ for some group Γ of type (g,n).

THEOREM 9.1 (Bers [10]). *If* $2g + n > 2$, *then* $T(g,n+1)$ *is the holomorphic universal covering space of* $V(g,n)'$.

If we take Γ to be of type (g,n) without parabolic elements, then we may define a another fiber space $V(g,n)$ (= the (unpunctured) <u>Teichmüller curve</u> $= F(\Gamma)/\Gamma$ for Γ as above) that depends only on the type of the group Γ chosen. Since each "puncture" on $V(g,n)'$ provides a holomorphic section

$$e_j : T(g,n) \to V(g,n)$$

of the projection π, we have constructed (see Earle [15]) a marked n-pointed family of closed Riemann surfaces of genus g. (Let g,n be a pair of integers satisfying $2g - 2+n > 0$. By an n-pointed family of closed Riemann surfaces of genus g, we mean a pair of complex manifolds V and B, a holomorphic map $\rho : V \to B$, n holomorphic sections $e_j : B \to V$ (that is, $\rho \circ e_j = id$) such that

i) ρ is a proper submersion,

ii) $\rho^{-1}(t)$ is diffeomorphic to a closed surface X of genus g, for all $t \in B$, and

iii) the sections $e_1, \ldots, e_n$ are disjoint (that is, $e_j(t) \neq e_k(t)$ for all $t \in B$ if $j \neq k$).

If we set

$$V' = V \setminus \bigcup_{j=1}^{n} e_j(B),$$

then $\rho : V' \to B$ is a smooth fiber bundle with fiber X_n (= $X \setminus \{$n distinct "distinguished" points$\}$) and structure group $\mathrm{Diff}^+(X_n)$ (= group of orientation preserving diffeomorphisms of X keeping the n-"distinguished" points fixed). If the structure group of this bundle is reduced to $\mathrm{Diff}_o^+(X_n)$, the path component of the identity in $\mathrm{Diff}^+(X_n)$, the family is said to be marked. A map of marked n-pointed families is a pair of holomorphic maps $F : V_1 \to V_2$ and $f : B_1 \to B_2$ such that $F(V_1') = V_2'$ and $(F|V_1', f)$ is a map of $\mathrm{Diff}_o^+(X_n)$-bundles.)

The fact that the complex structure of $T(g,n)$ is canonical is also contained in

THEOREM 9.2 (Grothendieck [20], Earle [15]). Given any marked n-pointed family of closed Riemann surfaces of genus g $\rho : V \to B$, there is a unique map of marked families

$$\begin{array}{ccc} V & \xrightarrow{F} & V(g,n) \\ \rho \downarrow & & \downarrow \pi \\ B & \xrightarrow{f} & T(g,n) \end{array} .$$

The universal property described in the above theorem uniquely determines both $V(g,n)$ and $T(g,n)$.

We conclude this chapter with two remarks: (1) The development of this section has points in common with the differential-geometric approach to Teichmüller space theory due to Earle-Eells [16], which is not covered in these lectures. (2) The problem of describing all holomorphic sections of the bundle $\pi : V(g,n) \to T(g,n)$ has been solved (in Hubbard [21] and Earle-Kra [17], for special cases, and in a forthcoming paper by the last two authors in the general case).

REFERENCES

[1] W. Abikoff, Constructability and Bers stability of Kleinian groups (to appear).

[2] L. V. Ahlfors, On quasiconformal mappings, J. Analyse Math., 3 (1953/4), 1-58.

[3] L. V. Ahlfors, Finitely generated Kleinian groups, Amer. J. Math., 86 (1964), 413-429 and 87 (1965), 759.

[4] L. V. Ahlfors, and L. Bers, Riemann's mapping theorem for variable metrics, Ann. of Math., 72 (1960), 385-404.

[5] A. Andreotti and A. L. Mayer, On period relations for abêlian integrals on algebraic curves, Annali Sc. Nor. Sup. Pisa, 21 (1967), 189-238.

[6] L. Bers, Quasiconformal mappings and Teichmüller's theorem in Analytic Functions, pp. 89-119, Princeton University Press, Princeton, New Jersey, 1960.

[7] L. Bers, Holomorphic differentials as functions of moduli, Bull. Amer. Math. Soc., 67 (1961), 206-210.

[8] L. Bers, A non-standard integral equation with applications to quasiconformal mappings, Acta Math., 116 (1966), 113-134.

[9] L. Bers, On boundaries of Teichmüller spaces and on Kleinian groups: I, Ann. of Math., 91 (1970), 570-600.

[10] L. Bers, Fiber spaces over Teichmüller spaces, Acta Math., 130 (1973), 89-126.

[11] L. Bers, On moduli of Kleinian groups (to appear).

[12] L. Bers and L. Ehrenpreiss, Holomorphic convexity of Teichmüller spaces, Bull. Amer. Math. Soc., 70 (1964), 761-764.

[13] L. Bers and L. Greenberg, Isomorphisms between Teichmüller spaces, Advances in the theory of Riemann surfaces, Ann. of Math. Studies, 66 (1971), 53-79.

[14] H. Cartan, Quotient d'un espace analytique par un group d'automorphismes, in Algebraic Geometry and Topology, pp. 90-102, Princeton University Press, Princeton, New Jersey, 1957.

[15] C. J. Earle, On holomorphic families of pointed Riemann surfaces, Bull. Amer. Math. Soc., 79 (1973), 163-166.

[16] C. J. Earle and J. Eells, A fibre bundle description of Teichmüller theory, J. Diff. Geom., 3 (1969), 19-43.

[17] C. J. Earle and I. Kra, On holomorphic mappings between Teichmüller spaces (to appear).

[18] H. M. Farkas and H. E. Rauch, Period relations of Schottky type on Riemann surfaces, Ann. of Math., 92 (1970), 434-461.

[19] F. Gardiner and I. Kra, Quasiconformal stability of Kleinian groups, Indiana University Math. J., 21 (1972), 1037-1059.

[20] A. Grothendieck, Techniques de construction en géométrie analytique, Sem. Cartan 60/61, expose 17 (with appendix by J. P. Serre, Rigiditi' de foncteur de Jacobi d'echelon $n \geq 3$).

[21] J. Hubbard, Sur la non-existence de sections analytiques a la courbe universelle de Teichmüller, C. R. Acad. Sc. Paris, 274 (1972), A 978-A979.

[22] I. Kra, On spaces of Kleinian groups, Comment. Math. Helv., 47 (1972), 53-69.

[23] A. Marden, On homotopic mappings of Riemann surfaces, Ann. of Math., 90 (1969), 1-8.

[24] A. Marden, The geometry of finitely generated Kleinian groups, Ann. of Math., (1974) (to appear).

[25] B. Maskit, Self-maps of Kleinian groups, Amer. J. Math., 93 (1971), 840-856.

[26] Z. Nehari, Schwarzian derivatives and schlicht functions, Bull. Amer. Math. Soc., 55 (1949), 545-551.

[27] H. E. Rauch, A transcendental view of the space of algebraic Riemann surfaces, Bull. Amer. Math. Soc., 71 (1965), 1-39.

5. METRICS ON TEICHMULLER SPACE*

H. L. Royden
Stanford University

In this chapter we discuss invariant metrics on the Teichmüller space $T(\Gamma)$ of a Fuchsian group Γ operating on the upper half plane U, whose elements have no fixed points in U and whose quotient space $W = U/\Gamma$ is a compact Riemann surface of genus g.

§1. TEICHMÜLLER SPACE AND ITS TANGENT SPACE

We let $M(\Gamma) = M(W)$ be the collection of Beltrami coefficients for Γ (see Kra, §1), and for each $\mu \in M(\Gamma)$ we let w_μ be the normalized homeomorphism of $\hat{\mathbb{C}}$ onto itself with Beltrami coefficient μ. Denote by Γ_μ the Fuchsian group $w_\mu \circ \Gamma \circ (w_\mu)^{-1}$. Then U/Γ_μ is again a compact Riemann surface W_μ which is obtained by defining a new complex structure on W by taking as a conformal atlas the restrictions of w_μ which are homeomorphisms of open sets of W into $\mathbb{C}$. Two elements μ and ν of $M(\Gamma)$ are said to be <u>equivalent</u> if $w_\mu \circ \gamma \circ (w_\mu)^{-1} = w_\nu \circ \gamma \circ (w_\nu)^{-1}$ for each $\gamma \in \Gamma$. This is the same as requiring $w_\mu = w_\nu$ on the real axis. It is not difficult to show (see Ahlfors [3]) that μ and ν are equivalent if and only if there is a conformal homeomorphism $\varphi : W_\mu \to W_\nu$ which is homotopic to the identity map on W. A Beltrami coefficient μ is said to be (globally) <u>trivial</u> if it is equivalent to 0. The Teichmüller space $T_g = T(\Gamma) = T(W)$ of Γ (or W) is the space of equivalence classes in $M(\Gamma)$. If we use the group structure on the space $M_0(\Gamma)$ of trivial Beltrami coefficients introduced in Chapter 4, then $T(\Gamma) = M(\Gamma)/M_0(\Gamma)$.

We have described the Teichmüller space T_g in terms of a fixed group $\Gamma = \Gamma_0$ which serves as a base point in T_g. However, we could just as well have started

*This research is supported in part by the National Science Foundation under grant GP 33942 X1.

with any other Fuchsian group Γ' with U/Γ' a compact surface of genus g, and we would find a natural biholomorphic equivalence between $T(\Gamma)$ and $T(\Gamma')$. Thus if we wish to consider properties of T_g in the neighborhood of a given point, $x \in T_g$, we may always choose our base group Γ so that x is the equivalence class of trivial Beltrami coefficients.

In order to introduce metrics in differential-geometric terms we need to characterize the tangent space of T_g. Since $T_g = M(\Gamma)/M_0(\Gamma)$, its tangent space will be the quotient of the tangent space of $M(\Gamma)$ modulo those tangent vectors which are tangent to the subspace $M_0(\Gamma)$. In our present context the tangent space to $M(\Gamma)$ is the Banach space of bounded measurable Beltrami differentials which are compatible with Γ, or, equivalently, the space of bounded measurable Beltrami differentials on the Riemann surface W. A curve through the origin in $M_0(\Gamma)$ is a family $\mu(t)$ of Beltrami coefficients belonging to a continuous family $w_{\mu(t)}$ of homeomorphisms of U onto itself. The derivative with respect to t of $w_{\mu(t)}$ at $t = 0$ is a vector field h compatible with Γ, i.e., an h such that $h(z) = h(\gamma(z))/\gamma'(z)$ for each $\gamma \in \Gamma$. The derivative of $\mu(t)$ with respect to t at $t = 0$ is $\partial h/\partial\bar{z}$. Thus the tangent space at 0 to M_0 is the set of Beltrami differentials μ such that $\mu = \partial h/\partial\bar{z}$ for some vector field h compatible with Γ, or, equivalently, of some vector field on W.

Such a Beltrami coefficient is called <u>infinitesimally trivial</u>. It is not difficult to show (see Ahlfors [3]) that a Beltrami differential μ is infinitesimally trivial if and only if $\int_W \mu Q = 0$ for each holomorphic quadratic differential Q on W. The space $\underset{\sim}{Q}(W)$ of quadratic differentials on W is just the space $B_2(\Gamma)$ consisting of automorphic forms of weight -4, defined in Chapter 3. It has dimension $3g - 3$, and we see again that T_g has dimension $3g - 3$.

Since we have a natural pairing

$$[\mu, Q] = \int_W \mu Q$$

between $M(W)$ and $\underset{\sim}{Q}(W)$, whose kernel in $M(W)$ is $M_0(W)$ and whose kernel in $\underset{\sim}{Q}(W)$ is 0, we see that $\underset{\sim}{Q}(W)$ is just the cotangent space to T_g at $\mu = 0$.

§2. DIFFERENTIAL METRICS ON T_g

In order to define a differential metric in T_g we must introduce a norm $\|\xi\|_x = F(x,\xi)$ on the vectors ξ in the tangent space at x. Thus F is a non-negative real function defined on the tangent bundle of T_g. If $x(t)$ gives a smooth curve in T_g, then the length of that curve is given by $\int F[x(t),\dot{x}(t)]dt$, and we may define the distance between two points as the infimum of the lengths of curves joining them. Thus different choices of F give rise to different metrics on T_g.

One can define a norm on the tangent space to T_g at W by defining a pseudo-norm on the tangent space to $M(W)$, i.e., on the space of Beltrami differentials on W, which annihilates the infinitesimally trivial ones.

Another procedure is to define a norm $G(\eta)$ in the cotangent space $\underset{\approx}{Q}(W)$ and obtain F by duality:

$$F(\xi) = \sup_{\eta} [\xi,\eta]$$

as η runs over all elements of $\underset{\approx}{Q}(W)$ with $G(\eta) = 1$. This method is particularly convenient, since $\underset{\approx}{Q}(W)$ is a concrete finite dimensional space.

§3. HERMITIAN METRICS

If the norms F or G come from a positive definite Hermitian inner product, we get a Hermitian metric for T_g. On $\underset{\approx}{Q}(W)$ we get such an inner product whenever we specify a conformal metric $\lambda|dz|$ on W by setting

$$(Q_1,Q_2) = \int_W Q_1(z)\overline{Q_2(z)}\,\lambda^{-2}(z)dx\,dy .$$

If we take λ to be the Poincaré metric given on U by $\lambda = 1/2y$, then the induced Hermitian metric on T_g is the Weil-Petersson metric.

If we are given an inner product $(\, , \,)$ on $\underset{\approx}{Q}(W)$ we can construct an orthonormal basis $Q_1,\ldots,Q_n$ for $\underset{\approx}{Q}$ and define a kernel of the Bergman type by setting

$$K(z,\bar{\zeta}) = \sum Q_i(z)\overline{Q_i(\zeta)} .$$

Then K is a holomorphic quadratic differential in the variable z and the conjugate of a holomorphic quadratic differential in ζ. It is not difficult to show (see Royden [10]) that the dual inner product in the tangent space is given by

$$(*) \qquad (\mu,\nu) = \int_W \int_W \mu(z)K(z,\bar{\zeta})\overline{\nu(\zeta)}\, dx\, dy\, d\xi\, d\eta \,.$$

Thus every Hermitian metric on T_g may be obtained by specifying a suitable Hermitian symmetric form $K(z,\bar{\zeta})$ which is a holomorphic quadratic differential in z and defining (μ,ν) by $(*)$. In the case of the Weil-Petersson metric the form K is given (on a fundamental region in U) by

$$K(z,\bar{\zeta}) = \frac{3}{2\pi} \sum_{\gamma\in\Gamma} \frac{\overline{\gamma'(\zeta)}^2 dz^2\, \overline{d\zeta}^2}{[z-\overline{\gamma(\zeta)}]^4} \,.$$

A naturally defined Hermitian pseudo-metric on T_g is obtained by taking an orthonormal basis $w_1,\ldots,w_g$ for the Abelian differentials on W and constructing the Bergman kernel form $k(z,\bar{\zeta}) = \sum w_i(z)\overline{w_i(\zeta)}$. Then the form $K = k^2$ is a Hermitian kernel which is a holomorphic quadratic differential in z. It follows from the positive definiteness of k that $(*)$ defines a Hermitian pseudo-metric on the tangent space of T_g at W, and this will be a metric (rather than a pseudo-metric) if and only if the products of the Abelian differentials on W span $\underset{\sim}{Q}(W)$. Thus this procedure gives us a Hermitian metric on T_g except on the hyper-elliptic locus, where the metric fails to be definite. It turns out (see Royden [10]) that this metric is the pull-back to T_g of the Bergman-Siegel metric in the Siegel upper half plane under the Riemann mapping which maps each W_μ in T_g to its Riemann period matrix.

§4. THE TEICHMÜLLER METRIC

There is a natural Finslerian metric on T_g which is the differential form of a metric considered by Teichmüller. Teichmüller defined a global metric on T_g by defining the distance between W and W_μ to be

$$d_T(W,W_\mu) = \frac{1}{2} \log \frac{1+k}{1-k} \,,$$

where $k = \inf \|\nu\|$ as ν ranges over all ν equivalent to μ. The infinitesimal form of this metric comes from the norm on the tangent space to T_g at W obtained by

$$\|\mu\|_T = \inf \|\mu-\nu\| ,$$

where $\| \ \|$ is the L^∞ norm and ν ranges over all infinitesimally trivial Beltrami differentials. Thus the Teichmüller metric is the quotient metric which T_g inherits from the L^∞ metric on $M(W)$ by considering T_g as $M(W)/M_0(W)$.

The Ahlfors-Bers-Teichmüller theorem (Ahlfors [2], Bers [4]) asserts that for a given μ the ν equivalent to μ which minimizes $\|\nu\|$ is always a Teichmüller differential, i.e., a Beltrami differential of the form $\nu = k\bar{\varphi}/|\varphi|$ where $\varphi \in \underset{\sim}{Q}(W)$, and that conversely such a differential minimizes $\|\nu\|$ among all equivalent Beltrami differentials. From this it follows that the Teichmüller metric is the integrated form of the differential metric introduced in the previous paragraph and that the curve given by $\mu(t) = t\bar{\varphi}/|\varphi|$, with $t \in [0,k]$, is a geodesic of length $d_T(W_0,W_\nu)$ joining W_0 and W_ν with $\nu = k\bar{\varphi}/|\varphi|$.

Since each open geodesic arc of T_g is of the form $t\bar{\varphi}/|\varphi|$ with $t \in [0,k)$, it can be completed to a closed arc by adding the endpoint $k\bar{\varphi}/|\varphi|$. It follows from the Hopf-Rinow-Myers theorem that T_g is complete in the Teichmüller metric.

This metric can also be defined by specifying the dual norm in the cotangent space to be

$$\|Q\| = \int_W |Q| .$$

One can analyze the smoothness of this infinitesimal metric on the cotangent bundle of T_g, and it turns out (see Royden [8]) that it is of class C^1 but not C^2. In fact, a careful analysis of the smoothness properties of this metric shows that the unit balls in the cotangent spaces at W_μ and W_ν are affinely equivalent only if there is a conformal mapping between W_μ and W_ν.

Thus the only isometries of T_g for the Teichmüller metric are those which take each equivalence class $[W_\mu]$ into an equivalence class $[W_\nu]$ consisting of Riemann surfaces conformally equivalent to W_μ.

Let H be the group of all quasiconformal homeomorphisms of W onto itself, or equivalently of all quasiconformal homeomorphisms h of the upper half plane U onto itself such that $h \circ \gamma \circ h^{-1} \in \Gamma$ for each $\gamma \in \Gamma$, and H_0 the subgroup of those homeomorphisms which are homotopic to the identity on W, or, equivalently, those homeomorphisms of U onto itself for which $h \circ \gamma \circ h^{-1} = \gamma$ for all $\gamma \in \Gamma$. Then the Teichmüller modular group or mapping class group $\mathrm{Mod}(W) = \mathrm{Mod}(\Gamma) = M(g,0)$ of W is defined to be the quotient group H/H_0. Then, as in §5 of Chapter 4, we have a homeomorphism of $\mathrm{Mod}(W)$ into the group of biholomorphic self-mappings of T_g onto itself defined as follows: For each element $\theta \in \mathrm{Mod}(W)$, let $h \in H$ be a quasiconformal homeomorphism in the coset θ and define $h^* : M(W) \longrightarrow M(W)$ by taking $h^*(\mu)$ to be the Beltrami coefficient of $w_\mu \circ h^{-1}$. Then $h^*(\mu)$ depends only on θ and on the equivalence class of μ in T_g. Thus h^* induces a mapping θ^* of T_g onto itself. This mapping θ^* is not only a biholomorphic self-mapping but is an isometry under the Teichmüller metric. It follows from the statement in the preceding paragraph that, conversely, each isometry of T_g with the Teichmüller metric is given by such a self-mapping which comes from the action of an element of the modular group.

§5. THE KOBAYASHI METRIC

Every complex manifold M has a natural pseudo-metric on it which is invariant under biholomorphic self-mappings and such that each holomorphic map from one complex manifold to another is distance-decreasing in this pseudo-metric. This metric was first defined and studied by Kobayashi [6], [7]. For a bounded domain in $\mathbb{C}^n$ this pseudo-metric is always a metric, and Wu [11] has shown that, if this metric is complete for a domain in $\mathbb{C}^n$, then the domain must be a domain of holomorphy. It can be shown (Royden [9]) that this metric can be defined by a differential metric as follows: Let $\langle x,\xi\rangle$ be an element in the tangent bundle to M with ξ a tangent vector at the point $x \in M$. Define

$$F(x,\xi) = \inf R^{-1} ,$$

where R ranges over the set of radii of disks in $\mathbb{C}$ centered at 0 which can be mapped into M by a holomorphic map f which takes 0 to x and which takes the unit tangent vector of the disk to ξ.

The map f of the unit disk into T_g given by $\zeta \longmapsto \zeta\bar{\varphi}/|\varphi|$ for a given $\varphi \in \underset{\sim}{Q}(W)$ is totally geodesic in the Teichmüller metric, and every geodesic in the Teichmüller metric lies in the image of such a map. Thus the pull-back of the

Teichmüller metric under such a map is the Poincaré metric for the disk and has Gaussian curvature -4 everywhere. This is the analogue for differential metrics of the condition for Kähler metrics of having holomorphic sectional curvature everywhere equal to -4. Just as for Kähler metrics with this property, we can show (Royden [8]) that C^1 differential metrics having this property also have the property that the pull-back of the metric by any holomorphic map of the disk into the manifold has Gaussian curvature at most -4. The Ahlfors version (Ahlfors [1]) of the Schwarz-Pick lemma then asserts that any holomorphic map of the disk into T_g is distance-decreasing from the Poincaré metric on the disk to the Teichmüller metric on T_g. Since the mappings considered at the beginning of the paragraph are isometric between the Poincaré and Teichmüller metric, and there is such a map in the direction of each tangent vector to T_g, it follows that the Teichmüller metric for T_g is also the Kobayashi metric for T_g. For further details see Royden [8].

Since the Kobayashi metric is invariant under biholomorphic self-maps, it follows that each biholomorphic self-map of T_g is an isometry in the Teichmüller metric. This shows that the only biholomorphic self-maps of T_g are those which arise through the action of the Teichmüller modular group Mod(W).

The result of Wu, together with the fact that T_g is equivalent to a bounded domain in $\mathbb{C}^n$, gives another proof of the fact that T_g is a domain of holomorphy. This was first shown be Bers and Ehrenpreis [5] using other methods.

REFERENCES

1. Ahlfors, L. V., An extension of Schwarz's lemma, Trans. Amer. Math. Soc., 43 (1938), 359-364.

2. ——————, On quasiconformal mappings, J. Analyse Math., 3 (1953-4), 1-58.

3. ——————, Lectures on quasiconformal mappings, Van Nostrand Math. Studies 10, Van Nostrand, Princeton, N. J., 1966.

4. Bers, L., Quasiconformal mappings and Teichmüller's theorem, in Analytic Functions by R. Nevanlinna et al., Princeton University Press, Princeton, N. J., 1960, 89-119.

5. ———— and L. Ehrenpreis, Holomorphic convexity of Teichmüller spaces, Bull. Amer. Math. Soc., 70 (1964), 761-764.

6. Kobayashi, S., Invariant distances on complex manifolds and holomorphic mappings, J. Math. Soc. Japan, 19 (1967), 460-480.

7. ——————, Hyperbolic manifolds and holomorphic mappings, M. Dekker, New York, 1970.

8. Royden, H. L., Automorphisms and isometries of Teichmüller space, Ann. of Math. Studies, 66 (1971), 369-383.

9. ——————, Remarks on the Kobayashi metric, Springer Lecture Notes in Math., 185 (1971), 125-137.

10. ——————, Invariant metrics on Teichmüller space, to appear.

11. Wu, H., Normal families of holomorphic mappings, Acta Math., 119 (1967), 193-223.

6. MODULI OF RIEMANN SURFACES

William Abikoff
Columbia University

A finitely generated Kleinian group G represents a finite union of hyperbolic Riemann surfaces of finite type. The theory of moduli is concerned with parametrizing families of Riemann surfaces (of finite type, in fact, primarily compact ones); one can use Kleinian groups as a tool in this study. The subject has roots in statements of Riemann and Poincaré, among others. The first systematic study was conducted by Fricke who constructed the space of marked Riemann surfaces. Later results were obtained using either algebraic (i.e. periods of abelian differentials) or differential geometric techniques. Among the practitioners of the former method are Torelli, Siegel, Satake, Baily [3], Mumford, Mayer, and Deligne and Mumford [8]. The latter technique, in its modern form, is based primarily on the penetrating insights of Teichmüller [12] and their subsequent interpretation and transformation into a coherent theory. This latter work is due to Ahlfors, Bers et al. Our discussion will focus on the differential geometric approach. In one respect it is more general than the algebraic approach since we may consider a surface with ramification points.

Let S be a marked Riemann surface of finite type with ramification points, together with a corresponding, possibly ramified, covering by the upper half plane U, and projection map π. We assume that the surface S has type (p,n) and signature $(p,n;\nu_1,\ldots,\nu_n)$ where the ν_i are either integers greater than one or the symbol ∞. In the former case the covering is ramified of degree ν_i at some z_i on S and in the latter case there is a puncture at z_i. The group of cover transformations associated to the covering is denoted by G and is a finitely generated Fuchsian group of the first kind.

We may construct many "moduli spaces." If we view S as a Riemannian 2-manifold with metric $ds^2 = \sigma(z)|dz|^2$, we may perform changes of conformal structure by perturbating the metric. At each point $z \in S$, a perturbed metric may be written as

$$ds_1^2 = |\sigma_1(z)|\,|dz + \mu(z)d\bar{z}|^2 \qquad \|\mu(z)\|_\infty < k < 1 .$$

Since we are only interested in conformal structure, the scale factor $\sigma_1(z)$ is irrelevant and the functions $\mu(z)$ in the open unit ball in $L_\infty(S)$ may be taken as a space of moduli for the given signature. The space constructed is the space of Beltrami coefficients $M(S)$ (see Royden's talk).

We may also lift $\mu(z)$ to a fundamental set for the action of G on U, extend it to all of U as a Beltrami coefficient for the group G, and set it equal to 0 on L, the lower half plane. The unique solution w to the Beltrami equation $w_{\bar{z}} = \mu w_z$, normalized so that $w(-i) = -i$ and $w'(-i) = 1$, conjugates G into a quasi-Fuchsian group G_μ. The set of such G_μ is another moduli space of S; it is the Teichmüller space $T(S)$. For fixed choice of the cover group G, Bers has given the following representation of $T(S)$:

$$T(G) = \{\{w,z\} \mid z \in L,\ w_{\bar{z}} = \mu w_z \text{ as above}\}$$

where $\{w,z\}$ is the Schwarzian derivative of w. $T(G)$ is a cell in the $3g - 3 + n$ dimensional complex vector space $B_2(G,L)$ of G-invariant bounded holomorphic quadratic differentials for G with support in L. $T(G)$ represents the quasiconformal deformations $S \to S'$ of the surface S as deformations $G \to G'$ of the group G, i.e. as isomorphisms of the group G into the Möbius group. There is an explicit normalization in the definition of the embedding so that $G' = G''$ exactly in the case that $S' = S''$ as marked surfaces. It is however possible that a deformation S' of S is conformally equivalent to S'' without G' being equal to G''. This phenomenon occurs when the conformal map $f: S'' \to S'$ does not preserve markings.

Another moduli space of a compact surface S without ramification, obtainable as a quotient of $T(G)$ by a discontinuous group, is the Torelli space $\mathcal{T}(S)$. It is the moduli space of the pair $\{S, \langle \beta_1, \ldots, \beta_n \rangle\}$ where the β's are a fixed basis of the first homology group of S. The Torelli space has been extensively studied by algebraic geometers and will not be considered here (see Kra's lecture).

The ultimate objects of our study are conformal equivalence classes of Riemann surfaces. If we restrict our attention to Riemann surfaces of fixed signature, every such surface is represented (infinitely often) in the Teichmüller space $T(S_0)$ of one surface S_0 of that signature. The relation of conformal equivalence in $T(S_0)$ is induced by a representation of a subgroup of the mapping class group $M_{p,n}$. The subgroup $M_{p,n}(S_0)$ of $M_{p,n}$ is the group of proper automorphisms of G_0, i.e. the group of changes of markings on S_0 which respect ramification orders. In some cases, the action of $M_{p,n}(S_0)$ on $T(S_0)$ is not effective. The Teichmüller modular group, already mentioned in Chapter 4, is defined to be exactly $M_{p,n}(S_0)$. Our definition differs in that $\text{Mod } G_0 = \text{Mod } S_0$ is $M_{p,n}(S_0)$ modulo its subgroup of ineffective elements. The arguments of Chapter 4 may then be repeated verbatim to show that the Riemann space of S_0,

$$R(G_0) = R(S_0) = T(G)/\text{Mod } G_0 ,$$

is a normal complex space. We have thus constructed a holomorphic parametrization of the conformal equivalence classes of Riemann surfaces of a given signature.

The theory of moduli of non-singular algebraic curves of genus one, i.e. compact surfaces of genus one without distinguished points is quite classical. It is known for example, that the field of meromorphic functions on such a surface is generated by the Weierstrass function $\wp(z;1,\tau)$ and $d\wp(z:1,\tau)/dz$ where $\tau \in U$ = Teichmüller space of Riemann surfaces of type (1,1).

For surfaces of higher genus the complex number τ is replaced by a point in $T(S_0)$. Among other results we have the following:

<u>Theorem 1</u>: <u>If</u> S_0 <u>is a hyperbolic Riemann surface of finite type with signature and</u> S_τ <u>is a surface corresponding to a point</u> τ <u>in</u> $T(S_0)$, <u>then the periods of normalized Abelian differentials on</u> S_τ <u>are holomorphic functions of</u> τ.

More striking results may be obtained using fiber spaces over the Teichmüller space (see Bers [5]).

Except in the most trivial of cases, the Riemann space $R(S)$ is not compact. This leads immediately to questions about possible "natural" compactifications of moduli spaces of S and their interpretations in terms of the surfaces S. $M(S)$ is much too large. The Satake compactification of the Torelli space (see Baily [3]) adds a "boundary" of too large a codimension. The Teichmüller space $T(S)$ as a bounded domain in a complex vector space has a natural compactification. But most boundary points correspond to degenerate groups (see Bers' lecture). Such groups "represent" the disappearance of the deformed surface S. Hence this compactification of itself yields virtually no information about limits of deformations of surfaces.

[Before proceeding, we note that "degeneracy" in the theory of Kleinain groups and in algebraic geometry has quite different meanings. In the following discussion of compactification of the space of moduli, the Kleinian groups used to represent degenerate algebraic curves are regular b-groups and constructible groups. These groups are not at all degenerate.]

We may consider the space of moduli of a hyperbolic surface S to be the space of conjugacy classes of Fuchsian groups of a fixed signature. The following recent generalization of a theorem of Mumford characterizes the compact subsets of $R(S)$.

Theorem 2 (Bers [6]): *Let G be a finitely generated Fuchsian group of the first kind, and let $X(G)$ be the space of conjugacy classes of Fuchsian groups isomorphic to G. Let $X_\epsilon(G)$ be the subset of $X(G)$ consisting of groups G' with $|\text{trace } \gamma| \geq 2 + \epsilon > 2$ for all hyperbolic $\gamma \in G$. $X_\epsilon(G)$ is compact.*

Geodesic curves on U/G' which are either simple loops or slits connecting ramification points of order 2, will be called *admissible curves*. The condition that trace γ be close to 2 for some $\gamma \in G'$ is equivalent to the existence of a short admissible curve on the surface U/G'. In particular, a divergent sequence of points in $R(S)$ represents a sequence of surfaces S_n containing admissible geodesic curves C_n whose lengths converge to zero. The compactification $\hat{R}(S)$ of $R(S)$ which we seek is one in which the points of $\hat{R}(S) - R(S)$ represent (topologically) surfaces obtained from S by the removal of a finite set of admissible curves. In terms of the conformal structures, the lengths of these admissible curves have been set equal to zero; we say that these curves have been *pinched*.

$\hat{R}(S)$ is called the *augumented moduli space*.

If S is a compact surface without distinguished points, Mumford and Mayer have proved, by algebraic methods,

Theorem 3: $\hat{R}(S)$ *is a compact normal complex space*.

There is no complete printed proof in the literature (cf., Deligne and Mumford [8] where the case of positive characteristic is treated).

The remainder of this talk is a discussion of work in progress on two analytic approaches to the problem of defining $\hat{R}(S)$ and determining its properties. There are other analytic approaches, e.g. using the space of Fuchsian groups (see Harvey [10]) and using techniques from 3-dimensional topology (see Marden's talk).

§1. PROPER PARTITIONS OF SURFACES AND SURFACES WITH NODES

1.1 Let S be a hyperbolic Riemann surface with signature and let $\{\alpha_1,\dots,\alpha_j\}$ be a set of disjoint admissible curves on S, such that $S\setminus\bigcup \alpha_i$ is a union of hyperbolic surfaces $S_1,\dots,S_k$ each of which is of finite type (p,n) with $3p-3+n \geq 0$. $P = \{S_1,\dots,S_k\}$ is called a *proper partition* of S and each element of P is called a *proper part* of S (in the proper partition P). It is well known that if S has genus p and m distinguished points, then

$$n \leq 3p - 3 - m \quad .$$

If equality holds, the partition P is called <u>maximal</u>. If P is maximal, then each S_j has type $(0,3)$; there are only finitely many homeomorphism classes of maximal partitions.

There is an obvious partial ordering on partitions and each partition may be refined to a maximal partition. It follows that there are finitely many homeomorphism classes of proper partitions.

The following theorem conjectured by Mumford was proved by Bers [7].

<u>Theorem 4</u>: <u>Every surface</u> S <u>with signature</u> $\sigma = (p,n;\nu_1,\ldots,\nu_n)$ <u>admits a maximal proper partition</u> P <u>defined by admissible curves whose lengths are bounded by a constant</u> L <u>depending only on</u> σ.

If we consider a sequence (S_n) of surfaces with the same signature, and partitions (P_n) induced by homeomorphisms $f_n: S_1 \to S_n$, then it is possible that the length of $f_n(\alpha_i) \to 0$, i.e. the sequence of deformations (S_n) of S_1 pinches the admissible curve α_i. Representations of such deformations by limits of quasiconformal deformations of Kleinian groups may be obtained in several ways. A sequence of quasi-Fuchsian groups may converge to a regular b-group in a "canonical" fashion (see Abikoff [ ]). The transformation pairing elliptic fixed points in the example given in Bers' talk (p. 9) may be deformed to the identity. There are other methods due to Maskit [11] and Harvey [10].

1.2 A Riemann surface S with nodes is a connected complex space such that every point $P \in S$ has a neighborhood isomorphic either to the unit disc (with P corresponding to the center) or to two unit discs, with the centers identified (and corresponding again to P). In the second case P is called a node. A part of S is a component of $S \setminus \{\text{all nodes}\}$. If we choose on S a discrete sequence of points, distinct from the nodes and assign to them ramification numbers (integers > 1 or ∞), we obtain a Riemann surface with nodes and ramification points. Every part of S is now a Riemann surface which is non-singular (i.e. without nodes) but with ramification points; we agree that a puncture produced by removing a node is to be considered a ramification point of order ∞. S is called of finite type if it has finitely many parts, each of finite type. S is called hyperbolic if each part is. We assume both conditions in what follows, and use on S the Poincaré metric defined on the parts.

A proper partition of S is defined just as before, except that the nodes are to be included among the admissible curves, and are assigned length 0. Theorem 4 remains valid. (The signature of S is now the set of signatures of the parts of S, together with a list of the pairs of ramification points of order ∞ which are joined in a node.)

A continuous surjection $f: S' \to S''$ is called a

deformation if $f(\text{node}) = \text{node}$, $f^{-1}(\text{node}) = $ node or Jordan curve avoiding nodes and ramification points, f(ramification point of order ν) = ramification point of order ν, f^{-1}(ramification point of order $\nu < \infty$) = ramification point of order ν, f^{-1}(ramification point of order ∞) = ramification point of order ∞ or a Jordan arc joining two ramification points of order 2 and avoiding all other ramification points and nodes, and if f, restricted to the complement of inverse images of nodes and Jordan arcs mapped into ramification points, is an orientation preserving homeomorphism.

If S is non-singular, $\hat{R}(S)$ can be defined, set-theoretically, as the set of isomorphism classes of all Riemann surfaces S' into which S may be deformed. If S is compact, without ramification points, and of genus $p > 1$, we denote $\hat{R}(S)$ by M_p.

§2. THE COMPACTIFICATION M_p

Now we describe an analytic construction of M_p. (See [7] and Bers' forthcoming paper for details and for more general constructions.)

We begin by choosing an integer $\nu > 3$ and $2p - 2$ Fuchsian groups $G_1, \ldots, G_{2p-2}$ of signature $(0,3;\nu,\nu,\nu)$ acting on disjoint discs $\Delta_1, \ldots, \Delta_{2p-2}$. The discs should be so far apart, that the G_j generate a Kleinian group representing $2p - 2$ surfaces of type $(0,3)$ and one surface Σ of type

$(0,6p - 6)$. Each G_j has in Δ_j a fundamental (non-Euclidean) 6-gon, 3 of whose vertices are elliptic fixed points. We number these $6p - 6$ points $\zeta_1,\ldots,\zeta_{6p-6}$, in such a way that $\Delta_1/G_1 \cup \ldots \cup \Delta_{2p-2}/G_{2p-2}$, with the images of ζ_{2j} and ζ_{2j-1} identified, becomes connected (and thus a Riemann surface S_1 with nodes); this can be done, of course, in different ways. We denote by ζ'_j the other fixed point of an elliptic element of order ν in G_j which keeps ζ_j fixed.

Let $t = (t_1,\ldots,t_{3p-3}) \in C^{3p-3}$ be such that $\zeta_{2j-1} + t_j \in \Delta_j$ for all j. Define $g_{t,j}$ to be the identity if $t_j = 0$, and the Möbius transformation which takes ζ_{2j-1}, ζ'_{2j-1} and $\zeta_{2j-1} + t_j$ into ζ'_{2j}, ζ_{2j} and $\zeta_{2j-1} + t_j$, respectively, if $t_j \neq 0$. Let G_t be the group generated by $G_1,\ldots,G_{2p-2}$ and $g_{t,1},\ldots,g_{t,3p-3}$ and let $\Xi \subset C^{3p-3}$ be the set of those t for which G_t is Kleinian and represents Σ and (after images of ζ_{2j-1} and ζ_{2j} are identified whenever possible) a Riemann surface $S_{1,t}$ with as many nodes as there are vanishing components t_j of t, and of Poincaré area $2\pi(2p-2)$. The origin is an interior point of Ξ and the component of the interior of Ξ containing the origin will be denoted by $X(S_1)$.

The domain $X(S_1)$ turns out to be a <u>cell</u>. Its points represent Riemann surfaces, with or without nodes, which can be deformed into S_1. Of course, distinct points of $X(S_1)$ can represent isomorphic surfaces. Thus we have an equivalence relation ρ on $X(S_1)$. This relation is not described by a discontinuous group of automorphisms. Nevertheless, $X(S_1)/\rho$ is a normal complex space.

The construction of S_1 involved a choice: an appropriate ordering of $3p - 3$ elliptic vertices. If we do it is all possible ways, we obtain a certain number $s = s(p)$ of non-homeomorphic Riemann surface $S_1,\ldots,S_s$, each with $3p - 3$ nodes and $2p - 2$ parts of type $(0,3)$. We also obtain s simply connected bounded domains $X(S_1),\ldots,X(S_s)$ in $\mathbb{C}^{3p-3}$ and s normal complex spaces $X(S_1)/\rho,\ldots,X(S_s)/\rho$. Since the points of these spaces are isomorphism classes of Riemann surfaces, we can form their union, and it turns out that

$$X(S_1)/\rho \cup \ldots \cup X(S_s)/\rho = M_p \quad .$$

Furthermore, the complex structures of the s space are compatible and M_p becomes a normal complex space.

Using Theorem 4 one can show that each $X(S_j)$ has a compact part K_j, such that $M_p \subset K_1/\rho \cup \ldots \cup K_s/\rho$. Hence M_p is compact.

§3. $\hat{R}(S)$ AS A QUOTIENT OF THE AUGMENTED TEICHMÜLLER SPACE

The augmented Teichmüller space $\hat{T}(S)$ or $\hat{T}(G)$ of a surface S or group G is the usual Teichmüller space together with the regular b-groups on its boundary. The Mumford-Mayer theorem suggests that some fundamental set for Mod S or Mod G

may be compactified by some of the regular b-groups in $\partial T(S)$ or $\partial T(G)$. We may thereby obtain a compactification of $\hat{R}(S)$ or $\hat{R}(G)$. Details will appear in the forthcoming paper of Abikoff.

We first define a topology on $\hat{T}$ which extends the usual topology on T. We must only define a neighborhood filter for a regular b-group G_0. If G_0 is a regular b-group, it represents the mirror image $\bar{S}$ of a surface S and (topologically) a proper partition of S. A neighborhood $N_{K,\epsilon}(G_0)$ is determined by a neighborhood of the paired nodes on $S'(G_0) = ((\Omega(G_0)/G_0) - \bar{S})$ and a positive number ϵ. A group G_1 lies in $N_{K,\epsilon}(G_0)$ if there is a $(1+\epsilon)$-quasiconformal mapping f of $S'(G_o)-K$ into $S'(G_1)$ such that: f admits a locally quasiconformal extension to $S'(G_0)$ which is either onto $S'(G_1)$ or is onto $S'(G_1)$ minus an admissible curve. In this topology, the action of the modular group on $\hat{T}$ is continuous. $N_{K,\epsilon}(G_0)$ is a horocyclic neighborhood of G_0, i.e.

(1) if $g \in \mathrm{Mod}(G)$ and $g(G_0) = G_0$ then

$g(N_{K,\epsilon}(G_0)) = N_{K,\epsilon}(G_0)$, and

(2) if $g(G_0) = G_1$ then $g(N_{K,\epsilon}(G_0))$ is a neighborhood of G_1

(3) if K,ϵ are sufficiently small, then

$$g(N_{K,\epsilon}(G_0)) \cap N_{K,\epsilon}(G_0) = N_{K,\epsilon}(G_0)$$

for g in the modular group.

It follows that $\hat{R}(\cdot) = \hat{T}(\cdot)/\mathrm{Mod}(\cdot)$ has a well defined topological space and is, in fact, Hausdorff. To show that it is compact, we must show that if $G_n \in \hat{T}(G)$ then there exists a subsequence G_{n_i} and a sequence $g_i \in \text{Mod } G$ so that $\{g_i(G_{n_i})\} \subset\subset T(G)$ or $g_i(G_{n_i}) \subset N_{K,\varepsilon}(G_0)$ for some regular b-group G_0. For any $H \in \hat{T}(G_0)$, a slight extension of the Mumford conjecture noted in §2 yields the existence of a proper partition P of $S'(G_1)$ defined by $3g - 3 + m$ admissible curves of uniformly bounded length. For a sequence $G_n \in \hat{T}(G_0)$ only finitely many of the corresponding partitions P_n may be inequivalent under $\text{Mod } G_0$. By passing to a subsequence we may find $g_i \in \text{Mod } G_0$ and groups G_{n_i} so that $g_i(G_{n_i})$ are convergent in $B_2(G_0)$ and the partitions of $S'(g_i(G_{n_i}))$ are consistent, i.e. the partitions P_{n_i} are defined by the same curves relative to a consistent marking of the $S'(G_{n_i})$. A slight generalization of Theorem 2 of Abikoff [1] says that the lengths of the admissible curves relative to a fixed marking and defining a proper partition, cannot stay uniformly bounded if the $g_i(G_{n_i})$ converge to a degenerate b-group. It follows that $g_i(G_{n_i})$ converge to a regular b-group or quasi-Fuchsian group in the relative topology of $\hat{T}(G_0)$ in $B_2(G_0)$. A more refined argument shows that the convergence actually occurs in the $\hat{T}$-topology defined above.

REFERENCES

[1] W. Abikoff, Two theorems on totally degenerate Kleinian groups, to appear.

[2] __________, to appear.

[3] W. L. Baily, Jr., On moduli of Jacobian varieties, Ann. of Math., 71(1960), 303-314.

[4] L. Bers, Holomorphic differentials as functions of moduli, Bull. Amer. Math. Soc., 67(1961), 206-210.

[5] ______, Fiber spaces over Teichmüller spaces, Acta Math., 130(1973), 89-126.

[6] ______, A remark on Mumford's compactness theorem, Israel J. of Math., 12(1972), 400-407.

[7] ______, Spaces of degenerate Riemann surfaces, Discontinuous Groups and Riemann Surfaces, Ann. of Math. Studies, 79(1974).

[8] P. Deligne and D. Mumford, Irreducibility of the space of curves of a given genus, I.H.E.S., 36(1969), 75-109.

[9] C. Earle and A. Marden, to appear.

[10] W. Harvey, to appear in Discontinuous Groups and Riemann Surfaces, Ann. of Math. Studies, 79(1974).

[11] B. Maskit, to appear.

[12] O. Teichmüller, Extremale quasikonforme Abbildungen und quadratische differentiale, Abh. Preuss. Akad. Wiss. Math. Nat. Kl., 22(1939).

7. GOOD AND BAD KLEINIAN GROUPS

Bernard Maskit

SUNY, Stony Brook

There are several well known characterizations of the class of finitely generated Fuchsian groups of the first kind. More generally, there is a class of good Kleinian groups with an invariant component which can be completely classified in exactly the same sense that these Fuchsian groups can be completely classified. This class of good Kleinian groups also has several different characterizations. These characterizations can be used to distinguish a class - or perhaps several classes - of good Kleinian groups which do not necessarily have an invariant component. Very little is known about such groups. Even less is known about Kleinian groups which are not good.

§1. FUCHSIAN GROUPS

1.1 The following theorem, which gives several of the many characterizations of a class of Fuchsian groups, is well known.

Theorem. _For a Fuchsian group_ Γ _acting on the upper half plane_ U, _the following statements are equivalent._

i) Γ is finitely generated and of the first kind.

ii) U/Γ is a finite Riemann surface; i.e. U/Γ is a compact Riemann surface from which finitely many points have been removed and the covering $U \to U/\Gamma$ is branched over finitely many points.

iii) U/Γ has finite non-Euclidean area.

iv) Γ has a finite-sided fundamental polygon, and is of the first kind.

1.2 One can (and sometimes should) think of Kleinian groups as being Fuchsian groups - operating on the 3-dimensional disc - of the second kind. There is also a class of "good" Fuchsian groups, which are not necessarily of the first kind; i.e., the finitely generated ones. The following theorem is also well known.

Theorem. For a Fuchsian group Γ acting on the upper half-plane U, the following statements are equivalent.

i) Γ is finitely generated.

ii) U/Γ is homeomorphic to the interior of a compact orientable 2-manifold and the covering $U \to U/\Gamma$ is branched over finitely many points.

iii) K/Γ has finite non-Euclidean area, where K is the Nielsen convex region; i.e. K is the non-euclidean convex hull of Λ (= limit set of Γ).

iv) Γ has a finite-sided fundamental polygon.

§2. CLASSIFICATION OF FUCHSIAN GROUPS

2.1 Finitely generated Fuchsian groups of the first kind have been classically classified; we describe one form of this classification.

There is a countable collection $\{\Gamma_i\}$ of finitely generated Fuchsian groups of the first kind with the following properties.

i) If Γ is a finitely generated Fuchsian group of the first kind, then for some i, Γ is a quasiconformal deformation of Γ_i.

ii) If $i \neq j$, Γ_i is not a quasiconformal deformation of Γ_j, in fact the coverings $U \to U/\Gamma_i$ and $U \to U/\Gamma_j$ are topologically distinct.

iii) Each point in the Teichmüller space of Γ_i (the space of quasiconformal deformations of Γ_i which keep U invariant and which are appropriately normalized) corresponds to a unique isomorphism of Γ_i onto some Fuchsian group; the mapping from the Teichmüller space onto such isomorphisms is one-to-one and real analytic (if one carefully chooses a subclass of allowable isomorphisms, then the mapping is also onto).

iv) One can topologically describe the covering $U \to U/\Gamma_i$ as the <u>branched</u> <u>universal</u> <u>covering</u> of U/Γ_i.

§3. CLASSIFICATION OF GOOD FUNCTION GROUPS

3.1 A *function group* is a finitely generated Kleinian group with an invariant component. A *good function group* is a function group which satisfies any one of the conditions (A), (C), (C'), (D), (E) or (F) given in the paragraphs below; an outline of the proof that these conditions are all equivalent appears in §12.

3.2 As for Fuchsian groups, there is a countable list of function groups. The list given in [7] contains some redundancies, but one easily normalizes further so as to get a non-redundant list of good function groups $\{G_i\}$, where G_i has invariant component Δ_i, with the following properties.

i) If G is a good function group with invariant component Δ, then G is a quasiconformal deformation of some G_i, where the quasiconformal deformation w maps Δ_i onto Δ.

ii) If $i \neq j$, G_i is not a quasiconformal deformation of G_j (there might be a homeomorphism $f : \Delta_i \to \Delta_j$ which conjugates G_i onto G_j, but this conjugation would not preserve parabolic elements in both directions).

iii) The deformation space (or space of quasiconformal deformations) $T(G_i)$ can be described as follows. $T(G_i)$ is a product of two factors $T'(G_i)$ and $T''(G_i)$. $T'(G_i)$ is the space of quasiconformal deformations with support in Δ_i; it is a manifold whose universal covering space is the Teichmüller space of the branched universal covering group of Δ_i/G_i. $T''(G_i)$ is the space of deformations whose support lies outside Δ_i, it is a product of lower dimensional Teichmüller spaces (see Kra's lecture).

iv) Each of the groups G_i can be uniquely described up to quasiconformal deformation in terms of the covering $\Delta_i \to \Delta_i/G_i$. Roughly, this description is as follows. Let S be a finite Riemann surface; let $w_1,\ldots,w_n$ be a set of simple disjoint homotopically distinct loops on S; and let $\alpha_1,\ldots,\alpha_n$ be positive integers (we also allow $\alpha_i = \infty$). The covering $\Delta_i \to \Delta_i/G_i$ is the highest regular covering of Δ_i/G_i for which the loops $w_j^{\alpha_j}$ $(\alpha_j < \infty)$ all lift to loops; the elements of G_i corresponding to those w_j where $\alpha_j = \infty$ are parabolic; and in a natural sense, the preceding statements account for all parabolic elements of G_i.

3.3 The statements in 3.2 do not appear as such in print, but are easy consequences of known results. Starting with statement iv) the groups G_i are constructed in [8] and [9]. Statement i) follows from Maskit [10] and Marden [6]. Statement ii) is simply normalization. The main part of statement iii) follows from Bers' technique of variation of parameters using quasiconformal mappings; the other statements are due to Kra [5] and Maskit [11].

§4. ALGEBRAIC CONDITIONS

4.1 When we try to separate good from bad groups, the most obvious algebraic condition is to require that G be finitely generated. One might expect to ask that G be finitely related, but there is a theorem of Scott [13] which implies that if G is finitely generated, then it is finitely related (see Marden's lecture).

One could view the statement that a particular element of G is parabolic as an algebraic statement, but even with this added information, we cannot separate good from bad groups with purely algebraic information. For example, there are degenerate groups - clearly bad - which are purely loxodromic and which are isomorphic to Fuchsian groups.

From here on we will deal only with groups which are finitely generated, for we expect that all our good groups will be.

§5. TOPOLOGICAL CONDITIONS

5.1 The most obvious topological condition comes from looking at the action of a Kleinian group G on the hyperbolic ball B^3 (see Marden's lecture); we could require that B^3/G be homeomorphic to the interior of a compact 3-manifold and that B^3 be branched over finitely many circles and points of B^3/G. One easily sees that if G satisfies this condition then G is finitely generated; nothing is known about the converse, nor is anything known about any relationship between this condition and questions concerning good and bad groups.

5.2 The next step is to look at the action of G on $\Omega(G)$ and to require that $\Omega(G)/G$ be a finite union of finite Riemann surfaces where Ω is branched over finitely many points. Ahlfors Finiteness Theorem [1] (also see Gardiner's lecture) asserts that if G is finitely generated then G satisfies this condition. The converse is false; furthermore both Fuchsian and degenerate groups satisfy this condition.

§6. METRIC CONDITIONS

6.1 We again regard G as acting on B^3, and obtain our first real condition.

(A) G has a finite-sided fundamental polyhedron.

A group satisfying condition (A) is called <u>geometrically finite</u>.

This condition can apparently be strengthened by requiring that <u>every</u> fundamental polyhedron have finitely many sides; it was shown by Beardon and Maskit [2] that if G satisfies (A), then indeed every convex fundamental polyhedron is finite-sided.

6.2 One can also form a generalized Nielsen convex region, K, and one could ask that K/G have finite volume. This condition presumably is equivalent to condition (A), but thus far nothing has been done along these lines.

§7. STABILITY

7.1 A Kleinian group G is called *stable* if every homomorphism from G into $PSL(2;\mathbb{C})$ which preserves parabolic elements, and which is sufficiently close to the identity, is in fact an isomorphism induced by a quasiconformal deformation with support in Ω.

We remark that it makes sense to ask if G is stable only if G is finitely generated.

(B) G is stable.

7.2 *Theorem (Marden [6]). If G is torsion free and G satisfies* (A), *then G satisfies* (B)

One expects the above to hold even if G is not torsion-free. The converse is not known, but it is known (Bers [3]) that degenerate groups are not stable; in fact, stability was introduced by Bers as part of his study of boundaries of Teichmüller spaces and degenerate groups.

One can also relate stability to certain cohomological conditions; see Gardiner and Kra [4].

§8. EXTENDABILITY OF MAPS

8.1 One possible criterion for niceness is that a group should be determined by its action on its set of discontinuity; i.e. for any Kleinian group G* other than G, if $\Omega(G^*)/G^*$ and $\Omega(G)/G$ "look alike" then G* should be a quasiconformal deformation of G.

In general, an isomorphism $\varphi : G \to G^*$ between Kleinian groups is called <u>type-preserving</u> if both φ and φ^{-1} preserve parabolic elements, and φ preserves the square of the trace of every elliptic element.

8.2 A group G is <u>quasiconformally extendable</u> if every type-preserving isomorphism $\varphi : G \to G^*$ which is induced by a quasi-conformal homeomorphism of $\Omega(G)$ onto $\Omega(G^*)$ is in fact induced by a quasiconformal deformation.

One expects every isomorphism induced by a homeomorphism of $\Omega(G)$ onto $\Omega(G^*)$ to be type preserving; this is known only for good function groups.

Similarly, a group G is <u>conformally extendable</u> if every type preserving isomorphism between G and some other group G*, which is induced by a conformal homeomorphism of $\Omega(G)$ onto $\Omega(G^*)$, is in fact induced by a fractional linear transformation.

(C) G is conformally extendable.

(C') G is quasiconformally extendable.

Using the existence of solutions of the Beltrami equation, one easily sees that if G satisfies (C), then G satisfies (C'); the converse is not known.

8.3 <u>Theorem (Marden [6])</u>. <u>If</u> G <u>is torsion-free and</u> G <u>satisfies</u> (A), <u>then</u> G <u>satisfies</u> (C).

The converse to this theorem is not known, nor are any relationships between conditions (B) and (C) known.

It was shown by Bers [3] that degenerate groups do not satisfy (C); one expects that they also do not satisfy (C'), but this is not known.

§9. THE LIMIT SET

9.1 One of the first possible conditions was that given by Ahlfors in [1] where he asks if the limit set of G necessarily has zero 2-dimensional measure. There are non-finitely generated groups whose limit sets have positive measure, but it is not known whether there are finitely generated groups with positive area limit sets.

If there were a group which satisfied (C') but not (C), then that group would have a limit set of positive area.

9.2 For our next condition, we need some definitions.

If H is a subgroup of G, a set $A \subset \hat{\mathbb{C}}$ is called <u>precisely invariant</u> under H if A is invariant under H, and $g(A) \cap A = \emptyset$, for all $g \in G-H$.

A fixed point z of a parabolic element of G is called <u>cusped</u> if G_z has rank 2, or if there is a set A which is the disjoint union of two open circular discs (or half-planes) where A is precisely invariant under G_z.

A limit point z of G is a <u>point of approximation</u> if there is a point x and there is a sequence $\{g_n\}$ of distinct elements of G so that the spherical distance $[g_n(z), g_n(x)]$ does not converge to 0.

(D) Every limit point of G is either a cusped parabolic fixed point or is a point of approximation.

9.3 Theorem (Beardon and Maskit [2]). G satisfies (A) if and only if G satisfies (D).

§10. MAXIMALITY

10.1 One of the most obvious characteristics of degenerate groups is that $\Omega(G)/G$ is in an obvious sense smaller than it should be. One possible way of describing this is in terms of dimensions of deformation spaces.

A Kleinian group G is *maximal* if for every type-preserving isomorphism ψ mapping G onto some other Kleinian group G*, the dimension of the deformation space of G is not less than that of G*.

(E) G is maximal.

Unfortunately, all that is known about this condition is that the usual bad groups don't satisfy this condition either.

There are several other maximality conditions which are equally well understood.

§11. CONSTRUCTIBILITY

11.1 There are two simple constructions which one can use to build more complicated groups from simpler ones; we outline these here, details appear in [12].

If H is a Fuchsian or quasi-Fuchsian subgroup of G, and B is an open topological disc which is precisely invariant under H in G, then we call B a <u>regular disc for</u> H if $\Omega(H) \cap \partial B \subset \Omega(G)$.

Combination I. Let γ be a simple closed curve which is invariant under the finitely generated Fuchsian or quasi-Fuchsian group H. Let the two open topological discs bounded by γ be B_1 and B_2. Suppose B_1 is a regular disc for H in G_1 and B_2 is a regular disc for H in G_2. Then G, the group generated by G_1 and G_2 is Kleinian and we say that G is formed from G_1 and G_2 via Combination I.

Combination II. Let γ_1 and γ_2 be disjoint simple closed curves bounding disjoint topological discs B_1 and B_2, where B_i is a smooth disc for H_i in G_1, $i = 1,2$; here H_i is a finitely generated Fuchsian or quasi-Fuchsian group. Suppose there is a transformation f which maps B_1 onto the complement of B_2 and conjugates H_1 into H_2. Then G, the group generated by G_1 and f is Kleinian, and we say that G is formed from G_1 and f via Combination II.

A group G is <u>constructible</u> if it can be built up from elementary groups and Fuchsian triangle groups using Combinations I and II a finite number of times. (This definition is not quite corect, complicated technical modifications are needed for groups with parabolic elements.)

(F) G is constructible.

One easily proves that if G satisfies (F), then G satisfies (D), and hence (A). One can also easily show that if G satisfies (F) then it satisfies (B). There are the usual examples of groups that do not satisfy (F).

§12. FUNCTION GROUPS

For function groups it was shown in [10] that (E) and (F) are equivalent. Using the Combination Theorem [12], one easily proves that (F) implies (D). We have already remarked that (D) is equivalent to (A) [2] which implies (B), (C) and (C') [6]. One shows that (C) (or (C')) implies (F) by using [10] and the non-uniqueness of degenerate groups due to Bers [3]. There is little doubt that (B) similarly implies (F), but there is no proof in the literature.

REFERENCES

1. L. V. Ahlfors, Finitely generated Kleinian groups, Amer. J. Math. 86 (1964), 413-429.

2. A. Beardon and B. Maskit, Limit points of Kleinian groups and finite-sided fundamental polyhedra, Acta Math. to appear.

3. L. Bers, On boundaries of Teichmüller spaces and on Kleinian groups: I, Ann. of Math. 91 (1970), 570-600.

4. F. Gardiner and I. Kra, Stability of Kleinian groups, Indiana Math. J. 21 (1972), 1037-1059.

5. I. Kra, On spaces of Kleinian groups, Comment. Math. Helv. 47 (1972), 53-69.

6. A. Marden, The Geometry of finitely generated Kleinian groups, Ann. of Math., to appear.

7. B. Maskit, Uniformizations of Riemann surfaces, to appear in Contributions to Analysis, Academic Press, New York, 1974.

8. ————, Construction of Kleinian groups, Proceedings of the Conference on Complex Analysis, Minneapolis, 1964, Springer-Verlag, Berlin, 1965, pp. 281-296.

9. ————, On Boundaries of Teichmüller spaces and on Kleinian groups: II, Annals of Math. 91 (1970), 607-639.

10. ————, Decomposition of certain Kleinian groups, Acta Math. 130 (1973), 243-263

11. ————, Self-maps on Kleinian groups, Amer. J. Math. 93 (1971), 840-856.

12. ————, On Klein's Combination Theorem, III, Advances in the Theory of Riemann Surfaces, Annals of Math. Studies 66 (1971), 297-316.

13. G. P. Scott, Finitely generated 3-manifolds are finitely presented, J. London Math. Soc. (2) 6 (1973), 437-440.

8. KLEINIAN GROUPS AND 3-DIMENSIONAL TOPOLOGY

A SURVEY

Albert Marden[1]
University of Minnesota
University of Maryland

§1. INTRODUCTION

In an 1883 paper in Acta Mathematica, Poincaré formulated the general theory of Kleinian groups. He based his approach on that which he had successfully applied to Fuchsian groups. The starting point was the recognition that a Möbius transformation acting in the plane can be regarded via stereographic projection as acting on the 2-sphere $\partial\mathcal{B}$ and then extended to the 3-ball $\mathcal{B}$ as well. Given a group G of Möbius transformations acting discontinuously on $\mathcal{B}$, Poincaré showed there were fundamental polyhedra in $\mathcal{B}$ for G exactly analogous to the fundamental polygons for Fuchsian groups. At this point however his general analysis ended.

In retrospect this is not surprising. For Fuchsian groups only involve surface topology and mathematicians have been dealing with this for a long time. In contrast, Kleinian groups involve 3-dimensional topology which is incredibly more complicated. Actually it is only relatively recently that a sufficient amount of information has been accumulated about

[1]**This work was supported in part by the National Science Foundation.**

3-manifolds for this theory to be an extremely useful tool in the study of Kleinian groups. The purpose of this report is to suggest why this is the case. It is based mainly on [7] to which the reader is referred for details and complete references.

§2. EXTENSION TO $\mathcal{B}$

Each Möbius transformation, acting on the sphere $\partial\mathcal{B}$, is the product of an even number of reflections in circles λ on $\partial\mathcal{B}$. Let σ be the sphere orthogonal to $\partial\mathcal{B}$ along the circle λ and set $\sigma_0 = \sigma \cap \mathcal{B}$. A reflection in λ can be extended to a reflection of $\mathcal{B}$ in σ_0. In this way the group of all Möbius transformations can be extended to act on $\mathcal{B}$. With respect to the hyperbolic metric $ds^2/(1-|x|^2)^2$ in $\mathcal{B}$, one gets the group of all orientation preserving isometries.

§3. DISCRETENESS AND DISCONTINUITY

A group G of Möbius transformations is discrete if it contains no sequence tending to the identity transformation. G is discrete if and only if it acts discontinuously in $\mathcal{B}$ (has no limit point in $\mathcal{B}$). Kleinian groups are always discrete but the converse is false: The limit set may be all of $\partial\mathcal{B}$ (sometimes these are called Kleinian groups of the first kind). One advantage of our approach is that it applies equally well to all discrete groups.

A theorem for 3-manifolds due to G.P. Scott [10] (independently proved by P. Shalen) implies that every finitely generated discrete group G is actually finitely presented.

So far, there is a reasonable theory only for finitely generated groups and we will stick to this case here. In addition we will always assume our groups are torsion free (contain no elliptic elements). Usually this is not an essential restriction because it follows from a result of A. Selberg that there is a torsion free subgroup of finite index. However in matters involving counting, for example of the sort involving Bers' area theorem, passing to a torsion free subgroup does not resolve the very difficult group-theoretic questions involved which in general have not been solved.

§4. THE 3-MANIFOLD

If G is discrete we can form the 3-manifold

$$\mathfrak{M}(G) = \mathbf{B} \cup \Omega(G)/G$$

which has a natural 3-dimensional conformal structure inherited from that of $\mathbf{B} \cup \Omega(G)$. The boundary

$$\partial\mathfrak{M}(G) = \Omega(G)/G$$

is a union of Riemann surfaces but is empty if $\Omega(G) = \emptyset$. In addition $\pi_1(\mathfrak{M}) \cong G$. (If G were allowed to have torsion it would still be true that $\mathfrak{M}(G)$ is a 3-manifold but no longer would $\pi_1(\mathfrak{M}) \cong G$.) If G is not elementary, each component of $\partial\mathfrak{M}(G)$ has the disk as universal cover. In particular no component is a sphere or a torus.

Examples. (1) If G is a finitely generated quasi-fuchsian group with Ω_0 one of its invariant components then

$\mathfrak{M}(G) \cong (\Omega_0/G)\times I$ (here $I = [0,1]$). For Fuchsian groups this can be seen directly.

(2) If G is a Schottky group of genus g, then $\mathfrak{M}(G)$ is a handlebody of genus g. That is $\mathfrak{M}(G)$ is homeomorphic to the compact region bounded by a surface of genus g embedded in $\mathbb{R}^3$.

(3) If G is a degenerate group it is unknown whether $\mathfrak{M}(G) \cong (\Omega(G)/G) \times [0,1)$. This fundamental problem will be discussed later in §12.

§5. FUNDAMENTAL POLYHEDRA

Given a point $0 \in \mathcal{B}$ the Poincaré fundamental polyhedron with center at 0 is defined as

$$\mathcal{P} = \{x \in \mathcal{B}: d(x,0) \leq d(x,T(0)), \quad \text{all} \quad T \in G\}.$$

Here $d(\cdot,\cdot)$ is the hyperbolic distance. Exactly as the analogous case for Fuchsian groups, the faces of $\mathcal{P}$ are arranged in pairs and the orbit of $\mathcal{P}$ under G covers $\mathcal{B}$ without overlapping interiors. With its opposite faces identified, $\mathcal{P}$ provides a model for $\mathfrak{M}(G)$.

§6. THE ROLE OF PARABOLIC TRANSFORMATIONS

Suppose $p \in \partial\mathcal{B}$ is a parabolic fixed point of G. The maximal parabolic subgroup at p is defined as

$$M_p = \{T \in G: T(p) = p\}.$$

It is known that M_p consists of parabolic transformations with a common fixed point and is either free abelian of rank two or infinite cyclic.

The role played in $\mathfrak{M}(G)$ by the rank two M_p's can be described precisely. There is a one-one correspondence (via lifting to $\mathcal{B}$) between conjugacy classes of these rank two M_p's and solid cusp tori in $\mathfrak{M}(G)$. A solid cusp torus is a submanifold $\mathfrak{T}$ in the interior $\mathfrak{M}(G)^o$, whose relative boundary in $\mathfrak{M}(G)^o$ is a torus, with

$$\mathfrak{T} \cong \{z \in \mathbb{C}: 0 < |z| < 1\} \times S^1.$$

The tori corresponding to distinct conjugacy classes can be taken to be mutually disjoint.

The role played in $\mathfrak{M}(G)$ by the cyclic M_p's is more complicated and is not completely understood. They are closely associated with punctures on $\partial\mathfrak{M}(G)$. A puncture is an ideal boundary component of a component of $\partial\mathfrak{M}(G)$ which has a neighborhood conformally equivalent to the once punctured disk. Corresponding to each puncture (via lifting to $\partial\mathcal{B}$) is the conjugacy class of a cyclic M_p. But there may be more than one puncture corresponding to a given class.

Two punctures (p,q) on $\partial\mathfrak{M}(G)$ are said to be paired if there is a submanifold $\mathfrak{C}$ of $\mathfrak{M}(G)$, which is called a solid cusp cylinder, with the properties that

$$\mathfrak{C} \cong \{z \in \mathbb{C}: 0 < |z| < 1\} \times [0,1]$$

and $\mathfrak{C} \cap \partial\mathfrak{M}(G)$ is a union of a neighborhood of p and one of q each of which is conformally equivalent to the once punctured disk. The relative boundary of $\mathfrak{C}$ in $\mathfrak{M}(G)^o$ is a cylinder. Furthermore if p and q are paired there is no third puncture paired with either p or q. The solid cusp cylinders corresponding to distinct pairs can be taken mutually disjoint.

A pair of punctures corresponds to the conjugacy class of a cyclic M_p. In general, however, this correspondence goes only in one direction.

§7. GROUPS OF COMPACT TYPE

This class is defined to consist of those non-elementary discrete groups with a finite sided Poincaré fundamental polyhedron. It is important because of the following fact [7]. The group G is of compact type if and only if $\mathfrak{M}(G)$ has the following structure. There are a finite number of mutually disjoint solid cusp cylinders and solid cusp tori so that their complement in $\mathfrak{M}(G)$ is compact. In particular if G has no parabolic transformations the condition is simply that $\mathfrak{M}(G)$ be compact.

One of the critical junctures in the theory of Kleinian groups was the discovery by Leon Greenberg [4] that degenerate groups are not of compact type.

On the other hand it follows from a result of Selberg and Garland-Raghunathan that if G has a fundamental polyhedron of finite hyperbolic volume (i.e. the coset space of G in $PSL(2,\mathbb{C})$ has finite volume) then G is of compact type. Recently Wielenberg [14] found an elementary proof of this in the context of Möbius groups and in fact proved a much stronger, local form of the result.

§ 8. THE CLASSIFICATION PROBLEM; THE COMPACT CASE

Suppose that $\mathfrak{M}(G)$ is compact and $\partial\mathfrak{M} \neq \emptyset$, in other words that G is a purely loxodromic Kleinian group of compact type. For technical reasons assume too that the inclusion $\pi_1(\partial\mathfrak{M}) \to \pi_1(\mathfrak{M})$ is injective, that is, that each component of $\Omega(G)$ is simply connected. According to Waldhausen [13], $\mathfrak{M}(G)$ has a hierarchy: One can successively introduce nondividing surfaces in $\mathfrak{M}(G)$ so that, cutting $\mathfrak{M}(G)$ along each of these surfaces in succession, after a finite number of steps $\mathfrak{M}(G)$ is reduced to a ball. Once this is accomplished the steps can be reversed thereby reforming $\mathfrak{M}(G)$ from a ball. With L. Greenberg and P. Scott we observed that this rebuilding process can actually be started with a handlebody of genus two (a Schottky group of genus two) rather than just a ball. The reconstruction can be described in terms of the Klein-Maskit combination theorems [8, 9].

In studying the application of the hierarchy to Kleinian groups one finds that a basic problem is to understand which of the topological operations one can use to form a new 3-manifold from some $\mathfrak{M}(G)$ can be carried out in the context of Kleinian groups (by using combination theory). For example consider a solid torus $\mathfrak{T}$ which arises from the action on $\mathfrak{B}$ by a cyclic group of loxodromic transformations. Take two disjoint disks in $\partial\mathfrak{T}$ and identify them to form a new 3-manifold. This can be carried out in the context of

Kleinian groups and one obtains a Schottky group of genus two. On the other hand suppose we take two disjoint, parallel annuli in $\partial\mathfrak{I}$ representing a non-trivial element in $\pi_1(\mathfrak{I})$. One can identify these to obtain a new 3-manifold but this process cannot be carried out in the context of Kleinian groups.

§9. DEFORMATION AND DEGENERATION THEORY

Given a group G of compact type what happens when one varies the entries in the matrices of a set of generators for G and uses these deformed matrices to generate a new group? More precisely, one wants to consider the space of all homomorphisms Hom G of G into $PSL(2,\mathbb{C})$ which preserve parabolic elements (actually a small modification of the Weil space $R(G,PSL(2,\mathbb{C}))$). Geometrically Hom G can be interpreted as a quasi-projective algebraic variety V(G). One is especially interested in the subset T(G) of V(G) consisting of discrete groups geometrically similar to G. It turns out that T(G) is a connected open subset of V(G) and in fact a complex analytic manifold. The proof [7] depends on proving the equivalence of the two topologies derived from a) the quasi-conformal deformations of G, and b) the topology of V(G). T(G) is the deformation or parameter space of G. From a different point of view the theory concerns conformal deformations of the conformal structure of $\mathfrak{M}(G)^{\circ}$, the topological structure remaining the same. This theory yields the result [7] that the dimension of T(G) (which after normalization is $\sum(3g_i+b_i-3)$ in the notation of §11) depends only on $\partial\mathfrak{M}(G)$, not on the internal structure of $\mathfrak{M}(G)$. If in particular $\partial\mathfrak{M}(G) = \emptyset$, then $T(G)$ is a point (this fact is an elaboration of Mostow's rigidity theorem).

As an open subset of the quasi-projective variety $V(G)$, $T(G)$ has a relative boundary $\partial T(G)$ in it. Chuckrow [3] proved that points on $\partial T(G)$ correspond to groups isomorphic to G and we noticed that these groups are in fact discrete although not necessarily Kleinian. Recently T. Jørgensen discovered a very general method of treating convergence of discrete groups and proofs of these results also follow immediately from the theorems of his elegant paper [5]. The points on $\partial T(G)$ which correspond to groups of compact type are called cusps because of certain analogies with the cusps of the classical modular group (for instance see Jørgensen's geometric description in [6]). Approaching a cusp has to do with pinching $\partial\mathfrak{M}(G)$ along one or more mutually disjoint simple loops. In the 3-manifold $\mathfrak{M}(G)$, a submanifold $\{\varepsilon < |z| < 1\} \times I$, where $\{\varepsilon < |z| < 1\} \times \partial I$ is the union of two annuli in $\partial\mathfrak{M}$, becomes a solid cusp cylinder (§6). In its general form, this degeneration theory is still in its infancy.

In the case of a Fuchsian surface group G, Bers has shown that $T(G)$ is esentially $T \times T$ where T is the ordinary Teichmüller space. The compactification of a slice T in $T(G) \cup \partial T(G)$ gives the Bers boundary ∂T of T. Using methods suggested by this general point of view one can extend the action of the Teichmüller modular group from T to the cusps on ∂T. This program is currently being carried out in joint work with Clifford Earle. It yields a compactification of the moduli space.

§10. THE CLASSIFICATION PROBLEM; THE NON-COMPACT CASE

Now we consider the case of an arbitrary finitely generated, torsion free Kleinian group G. The problem here is to find out how $\mathcal{M}(G)$ is related to a compact manifold. We will start by listing two questions.

(1) Does the interior $\mathcal{M}(G)^o$ of every $\mathcal{M}(G)$ contain a submanifold M with the properties

(i) the inclusion $\pi_1(M) \to \pi_1(\mathcal{M}(G))$ is an isomorphism, and

(ii) if N is a component of $\mathcal{M}(G)^o - M^o$ then the relative boundary ∂N of N in $\mathcal{M}(G)^o$ is connected, the inclusion $\pi_1(\partial N) \to \pi_1(N)$ is an isomorphism, and each lift of ∂N in $\mathbb{B}$ divides $\mathbb{B}$ into two components at least one of which is a ball?

(2) If G is a degenerate group and $S = \Omega(G)/G$, is $\mathcal{M}(G) \cong S \times [0,1)$?

We will discuss (2) in §12. The answer to (1) is affirmative when G is of compact type or is a function group. G.P. Scott [10] showed that the answer is affirmative and that M can be taken to be compact for all groups G such that G cannot be written as a free product of non-trivial subgroups.

He also showed that in the general case one can find a compact M satisfying (i).

The significance of these questions is this. Suppose G is a group for which (1) holds. Then in a certain sense one can completely understand how G is structurally made up of degenerate groups and groups of compact type. This is rather complicated to describe and we refer to [7] for details. The understanding of G and $\mathcal{M}(G)$ is such that if (2) were known to be true in addition, it would follow that $\mathcal{M}(G)^{\circ}$ is homeomorphic to the interior of a compact manifold.

§11. AHLFORS' FINITENESS THEOREM AND BERS' INEQUALITY

Consider a finitely generated Kleinian group G without elliptic elements. Let $\mathcal{M}'$ denote the manifold obtained from $\mathcal{M}(G)$ by removing the solid cusp tori. An elementary study [7] of the inclusion of first integral homology groups $H_1(\partial\mathcal{M}') \to H_1(\mathcal{M}')$ shows that

$$\sum g_i + c \leq N$$

where g_i is the genus of the ith component of $\partial\mathcal{M}(G)$, c is the number of solid cusp tori (or equivalently the number of conjugacy classes of rank two M_p's), and N is the number of generators of G. This formula should be compared with Bers' inequality [2]

$$\sum (g_i + b_i/2 - 1) \leq N - 1$$

where $b_i \geq 0$ is the number of punctures on the ith component of $\partial\mathcal{M}(G)$.

In order to investigate Bers' inequality further it is necessary to assume that $\mathcal{M}(G)$ has the property (1) of §10. Then it is possible to deduce a master inequality of which the two above are special cases. It shows for example that there is equality in Bers' inequality only in the following situation: Either $\mathcal{M}$ is compact with $\partial\mathcal{M}$ connected of genus N (Is G then a Schottky group?) or all the components of $\partial\mathcal{M}(G)$ have punctures and these are arranged in pairs so that when the corresponding solid cusp cylinders are removed from $\mathcal{M}(G)$, there remains a compact manifold $\mathcal{N}$ with $\partial\mathcal{N}$ a compact surface of genus N.

§12. THE FINAL EXAM

We believe that the most important unsolved problem in both Kleinian group and Teichmüller theory is the problem of determining the structure of degenerate groups G. The two aspects of this are finding the measure of the limit set and finding the topological structure of $\mathcal{M}(G)$. From the point of view of Kleinian groups this information is important in their classification (see §10) and from the point of view of Teichmüller theory it is necessary for a fuller understanding of the boundary.

Lest the reader be left with soaring enthusiasm in the power of 3-dimensional topology to solve problems in Kleinian groups, we close with the following important example of G.P. Scott based on work of Tucker [12] (see also [11]). Let S be a closed surface of genus ≥ 1. According to Scott (personal communication) there exists a 3-manifold M with the following properties.

(i) $\partial M = S$

(ii) The inclusion $\pi_1(S) \to \pi_1(M)$ is an isomorphism.

(iii) The universal covering space of M is the closed upper half space H.

(iv) If T is any cover transformation and $\langle T \rangle$ is the group generated by T then

$$H/\langle T \rangle \cong (S^1 \times \mathbb{R}) \times [0,1).$$

(v) $M \not\cong S \times [0,1)$.

Degenerate groups of course have properties (i) - (iv). But what about (v)? Consider the case that S is a torus. The corresponding thing for Mobius groups is the group G generated by say $z \mapsto z+1$, $z \mapsto z+i$. But here at least we know that $\mathfrak{M}(G) \cong S \times [0,1)$.

References

1. L.V. Ahlfors, Finitely generated Kleinian groups, Amer. J. Math. 86 (1964), 413-429 and 87 (1965), 759.

2. L. Bers, Inequalities for finitely generated Kleinian groups, Jour. d'Anal. Math. 18 (1967), 23-41.

3. V. Chuckrow, On Schottky groups with application to Kleinian groups, Ann. of Math. 88 (1968), 47-61.

4. L. Greenberg, Fundamental polyhedra for Kleinian groups, Ann. of Math. 84 (1966), 433-441.

5. T. Jørgensen, On discrete groups of Moebius transformation, to appear.

6. T. Jørgensen, On reopening of cusps, to appear.

7. A. Marden, The geometry of finitely generated Kleinian groups, Ann. of Math., 99 (1974).

8. B. Maskit, On Klein's combination theorem, Trans. A.M.S. 120 (1965), 499-509 and 131 (1968), 32-39.

9. B. Maskit, On Klein's combination theorem III, in Advances in the Theory of Riemann Surfaces, Annals of Math. Studies 66, Princeton University Press, Princeton, N.J.

10. G.P. Scott, Compact submanifolds of 3-manifolds, J. London Math. Soc. 7 (1973), 246-250.

11. G.P. Scott, An introduction to 3-manifolds, University of Maryland Lecture Notes.

12. T.W. Tucker, Some non-compact 3-manifold examples giving wild translations of $\mathbb{R}^3$, to appear.

13. F. Waldhausen, On irreducible 3-manifolds which are sufficiently large, Ann. of Math. 87 (1968), 56-88.

14. N. Wielenberg, On the fundamental polyhedra of discrete Moebius groups, Thesis, University of Minnesota, 1974.

THE CURVATURE OF TEICHMULLER SPACE

Howard Masur
University of Minnesota

(Abstract)

It is known that T_1, the Teichmüller space of a (punctured) torus, is isomorphic to the unit disc in the complex plane; in this case the Teichmüller metric is the Poincaré metric, and thus has negative curvature. For $g \geq 2$, the Teichmüller metric in the Teichmüller space T_g of a closed surface of genus g is not Riemannian. However Kravetz showed that T_g is straight, which means that between any two points there is a unique geodesic extending to infinity in either direction. A geodesic is an isometric image of the real line.

Busemann defined negative curvature in a straight space. Given any three non-collinear points P, Q, and R, let $\tilde{Q}$ be the midpoint of the segment $\overline{PQ}$ and $\tilde{R}$ the midpoint of $\overline{PR}$. Then the inequality $d(\tilde{Q},\tilde{R}) < \frac{1}{2} d(Q,R)$ is to hold for all choices of points P, Q, R. If a straight space has a Riemannian metric, this definition of negative curvature coincides with the usual one.

It was asserted that T_g, $g \geq 2$, has negative curvature. It is possible to prove however that T_g does not have negative curvature.

The idea is to find two geodesic rays r, s originating at the same point P, which satisfy $d(x,s) \leq M < \infty$ for $x \in r$, which is to say that r and s do not diverge. Then one sees that not all triangles with sides on r and s and vertex P can have the negative curvature property.

By the classical theorem of Teichmüller, the points on a ray through the point P, with underlying Riemann surface S, are determined by the Teichmüller extremal maps on S with dilatation $k\bar{\varphi}/|\varphi|$, where $0 \leq k < 1$ and φ is a holomorphic quadratic differential on S.

The crucial element in the proof is to find the "right" quadratic differentials on S. The quadratic differentials have closed horizontal trajectories and have been studied extensively by Strebel. Strebel's structure theorem says that S cut along the critical trajectories of such a quadratic differential is divided into annuli each equipped with a natural parameter. The Teichmüller map then can be described. It is simply an affine stretch of each annulus, with respect to these distinguished parameters, onto the image surface.

An existence theorem of Strebel's says that distinct differentials exist which divide the surface into annuli that are pairwise freely homotopic. These distinct differentials determine the required rays r and s. Using the fact that the Teichmüller map is easily described and that the annuli are pairwise freely homotopic, one shows that r and s do not diverge. Details will appear elsewhere.

SOME UNSOLVED PROBLEMS

Compiled by William Abikoff

The problems stated below were submitted by the lecturers shortly after the San Francisco meeting. The problems are, as far as we know, unsolved. No attempt has been made to indicate our sense of the relative difficulty of the problems or to give an exhaustive list of open problems. The classification of problems given below is more or less arbitrary. References to the literature are given with each problem.

I) Ahlfors' Zero Measure Problem:

If G is a finitely generated Kleinian group, is the two-dimensional measure of $\Lambda(G)$ equal to zero? (See Ahlfors [8], [9] and [11], Abikoff [1] and Kra's lecture.)

II) Teichmüller Spaces:

1) For a finitely or infinitely generated Fuchsian group G, is $T(G)$ starlike in the Bers embedding? (See Abikoff's lecture.)

2) Let G be a Fuchsian group and $T(G)$ its Teichmüller space. Is the Caratheodory metric on $T(G)$ the same as the Kobayashi metric? (See Royden's lecture and Earle [17].)

3) If G is a finitely generated Fuchsian group of the first kind, is there a natural notion of Kobayashi metric on

$\hat{T}(G)$ or on $\hat{R}(G)$? ($\hat{T}(G)$ and $\hat{R}(G)$ are defined in Abikoff's lecture, the Kobayashi metric and other relevant notions may be found in the lectures of Royden and Masur.)

4) If G is a fixed point free finitely generated Fuchsian group of the first kind, then the Bers fiber space is isomorphic to $T(G')$ for some G'. If G has elliptic elements, then in general no such isomorphism exists. For a finite number of cases the question is open. (See Kra's lecture; in particular, reference 17 of his lecture lists the open cases.)

5) Do all degenerate b-groups lie on the boundary of a finite dimensional Teichmüller space? (See Bers' lecture and Bers [15].)

6) If $f(z)$ is a schlicht function in the lower half-plane, does the Schwarzian derivative $\{f,z\}$ lie in the closure of $T(1)$, the universal Teichmüller space? (See Bers [13] and [15].)

7) Let G be a finitely generated Fuchsian group of the first kind and $T(G)$ be given the Bers embedding. Are groups with accidental parabolic transformations dense in the boundary of $T(G)$? (See Bers' lecture and Bers [15].)

8) Let G and $T(G)$ be as in Problem II-6. Mod G is a normal family of holomorphic automorphisms of the bounded domain $T(G)$. Let $g_n \in \text{Mod } G$ with $g = \lim g_n$. g is then a degenerate holomorphic mapping of $T(G)$ into $\partial T(G)$. Can g

be nonconstant without being onto a boundary Teichmüller space? (Boundary Teichmüller spaces are discussed in Abikoff [6]; see also Abikoff's lecture and his forthcoming paper.)

9) Let G and $T(G)$ be as in Problem II-6. Does there exist a fundamental set F for $\text{Mod}\ G$ so that if g_n is a sequence of elements of $\text{Mod}\ G$ converging to a degenerate holomorphic mapping g then either:

(i) $g(T(G))$ is a boundary Teichmüller space

or (ii) the diameters of $g_n(F)$ converge to zero?

(See the references for problem II-9)

10) Is the image of the canonical injection of $\hat{T}(G)$ into the affine algebraic variety of homomorphisms for the finitely generated Kleinian group G into Möb a submanifold? (See Kra's lecture for notation and background.)

11) Find or estimate the inradius of $T(G)$ in the Bers embedding. (See Chu [16] and Abikoff's lecture.)

III) <u>Automorphic Forms and Eichler Cohomology</u>

1) Is the canonical map $B_q(\Omega,\Gamma) \to H^1(\Gamma,\pi_{2q-2})$ injective for $g > 2$ and Γ infinitely generated? (See Gardiner's lecture for notation.)

2) For finitely generated Kleinian groups, the Eichler cohomology groups $H^1(\Gamma,\pi_{2q-2})$ can be decomposed into direct sums of cusp forms and quasi-bounded Eichler integrals. Is there such a decomposition for infinitely generated groups? (See Gardiner's lecture for description and references.)

3) What does the presence of trivial Eichler integrals tell us about the structure of the Kleinian group? (Again see Gardiner's lecture and the references given there.)

4) Let G be a Kleinian group. Are there any constructive characterizations of the kernel of the Poincaré theta operator? (See Bers [12] and Ahlfors [8].)

IV) Geometry and Topology of Kleinian Groups

1) Few properties of totally degenerate Kleinian groups are known, find more. (See Abikoff [5], Bers [15], Marden [18], Maskit [19] and [21].)

2) If one allows quasiconformal deformations to be supported on the limit set, are degenerate groups stable? (See Bers [14].)

3) Classify the finitely generated web groups. (See Abikoff [2] and [3] and Abikoff and Maskit [7].)

4) Which 3-manifolds are uniformizable by Kleinian groups, i.e. admit metrics of constant negative curvature? (See Marden's lecture.)

5) For a general (i.e. non-function) Kleinian group, are conditions (A), (B), (C'), (E) and (F) equivalent? (The conditions are stated in Maskit's lecture.)

6) For a Kleinian group G, let K be the region in B^3 defined as the intersection of all hyperbolic half-spaces whose boundaries lie in $\Omega(G)$. For which groups does $(B^3-K)/G$ have finite hyperbolic volume? (See Marden's lecture and Marden [18].)

7) Exactly as in the definition of conformally extendable, one can define the notion of topologically extendable. Does there exist a Kleinian group (not necessarily finitely generated) which is topologically extendable but not quasiconformally extendable? (See Maskit's lecture and Maskit [20] for the relevant notions.)

8) Do Maskit's combination theorems preserve Bers stability when the amalgamating subgroups or conjugated subgroups are of the second kind? (See Abikoff [4].)

9) Let G be a finitely generated Kleinian group of the first kind whose quotient has infinite hyperbolic volume. Does it necessarily follow that G has a finitely generated degenerate or non-constructible web subgroup? (Kleinian groups of the first kind are defined in Ahlfors [10], for the other notions see Problem IV-6 and the references given there and Abikoff [3].)

REFERENCES

[1] W. Abikoff, Some remarks on Kleinian groups, Advances in the Theory of Riemann Surfaces, Ann. of Math. Studies, 66(1970), pp.1-7.

[2] W. Abikoff, Residual limit sets of Kleinian groups, Acta Math., 130(1973), pp.127-144.

[3] W. Abikoff, On the decomposition and deformation of Kleinian groups, Contributions to Analysis, Academic Press, New York, 1974, pp.1-10.

[4] W. Abikoff, Constructibility and Bers stability of Kleinian groups, Discontinuous Groups and Riemann Surfaces, Ann. of Math. Studies, 79(1974), pp.3-12.

[5] W. Abikoff, Two theorems on totally degenerate Kleinian groups, to appear.

[6] W. Abikoff, On boundaries of Teichmüller spaces and on Kleinian groups, III, to appear.

[7] W. Abikoff and B. Maskit, to appear.

[8] L. Ahlfors, Finitely generated Kleinian groups, Amer. J. of Math., 86(1964), pp.413-429.

[9] L. Ahlfors, Some remarks on Kleinian groups, Proc. Tulane Conference on Quasiconformal Maps, Moduli and Discontinuous Groups, (1965), pp.7-13.

[10] L. Ahlfors, Kleinische Gruppen in der Ebene und im Raum, Festband 70. Geburtstag R. Nevanlinna, p.7-15, Springer, 1966.

[11] L. Ahlfors, Remarks on the limit point set of a finitely generated Kleinian group, Advances in the Theory of Riemann Surfaces, Ann. of Math. Studies, 66(1971), pp.19-26.

[12] L. Bers, Automorphic forms and Poincare series for infinitely generated Fuchsian groups, Amer. J. of Math., 87(1965), pp.196-214.

[13] L. Bers, Universal Teichmüller space, Analytic Methods in Mathematical Physics, Gordon and Breach, (1970), pp.65-83.

[14] L. Bers, Spaces of Kleinian groups, Lecture Notes in Math., 155(1970), Springer, Berlin, pp.9-34.

[15] L. Bers, On boundaries of Teichmüller spaces and on Kleinian groups, I, Ann. of Math., 91(1970), pp.570-600.

[16] T. Chu, On the outradius of finite-dimensional Teichmüller spaces, Discontinuous Groups and Riemann Surfaces, Ann. of Math. Studies, 79(1970), pp.75-80.

[17] C. Earle, On the Caratheodory metric in Teichmüller spaces, Discontinuous Groups and Riemann Surfaces, Ann. of Math. Studies, 79(1974), pp.99-104.

[18] A. Marden, Geometry of finitely generated Kleinian groups, to appear.

[19] B. Maskit, On boundaries of Teichmüller spaces and on Kleinian groups, II, Ann. of Math., 91(1970), pp.607-639.

[20] B. Maskit, Self-maps on Kleinian groups, Amer. J. of Math., 93(1971), pp.840-856.

[21] B. Maskit, to appear.

l. 247: Lectures on Operator Algebras. Tulane University Ring and perator Theory Year, 1970–1971. Volume II. XI, 786 pages. 1972. M 40,–

l. 248: Lectures on the Applications of Sheaves to Ring Theory. ane University Ring and Operator Theory Year, 1970–1971. Vol- ne III. VIII, 315 pages. 1971. DM 26,–

l. 249: Symposium on Algebraic Topology. Edited by P. J. Hilton. , 111 pages. 1971. DM 16,–

l. 250: B. Jónsson, Topics in Universal Algebra. VI, 220 pages. 72. DM 20,–

l. 251: The Theory of Arithmetic Functions. Edited by A. A. Gioia d D. L. Goldsmith VI, 287 pages. 1972. DM 24,–

l. 252: D. A. Stone, Stratified Polyhedra. IX, 193 pages. 1972. 18,–

l. 253: V. Komkov, Optimal Control Theory for the Damping of Vibra- ns of Simple Elastic Systems. V, 240 pages. 1972. DM 20,–

l. 254: C. U. Jensen, Les Foncteurs Dérivés de $\varprojlim$ et leurs Ap- cations en Théorie des Modules. V, 103 pages. 1972. DM 16,–

l. 255: Conference in Mathematical Logic – London '70. Edited by Hodges. VIII, 351 pages. 1972. DM 26,–

l. 256: C. A. Berenstein and M. A. Dostal, Analytically Uniform aces and their Applications to Convolution Equations. VII, 130 ges. 1972. DM 16,–

l. 257: R. B. Holmes, A Course on Optimization and Best Ap- oximation. VIII, 233 pages. 1972. DM 20,–

l. 258: Séminaire de Probabilités VI. Edited by P. A. Meyer. VI, 3 pages. 1972. DM 22,–

l. 259: N. Moulis, Structures de Fredholm sur les Variétés Hilber- nnes. V, 123 pages. 1972. DM 16,–

l. 260: R. Godement and H. Jacquet, Zeta Functions of Simple gebras. IX, 188 pages. 1972. DM 18,–

l. 261: A. Guichardet, Symmetric Hilbert Spaces and Related To- cs. V, 197 pages. 1972. DM 18,–

l. 262: H. G. Zimmer, Computational Problems, Methods, and sults in Algebraic Number Theory. V, 103 pages. 1972. DM 16,–

l. 263: T. Parthasarathy, Selection Theorems and their Applica- ns. VII, 101 pages. 1972. DM 16,–

. 264: W. Messing, The Crystals Associated to Barsotti-Tate oups: With Applications to Abelian Schemes. III, 190 pages. 72. DM 18,–

l. 265: N. Saavedra Rivano, Catégories Tannakiennes. II, 418 ges. 1972. DM 26,–

l. 266: Conference on Harmonic Analysis. Edited by D. Gulick d R. L. Lipsman. VI, 323 pages. 1972. DM 24,–

. 267: Numerische Lösung nichtlinearer partieller Differential- und egro-Differentialgleichungen. Herausgegeben von R. Ansorge und Törnig, VI, 339 Seiten. 1972. DM 26,–

. 268: C. G. Simader, On Dirichlet's Boundary Value Problem. IV, 8 pages. 1972. DM 20,–

. 269: Théorie des Topos et Cohomologie Etale des Schémas. GA 4). Dirigé par M. Artin, A. Grothendieck et J. L. Verdier. XIX, 5 pages. 1972. DM 50,–

. 270: Théorie des Topos et Cohomologie Etale des Schémas. me 2. (SGA 4). Dirigé par M. Artin, A. Grothendieck et J. L. Verdier. 418 pages. 1972. DM 50,–

. 271: J. P. May, The Geometry of Iterated Loop Spaces. IX, 175 ges. 1972. DM 18,–

272: K. R. Parthasarathy and K. Schmidt, Positive Definite Ker- s, Continuous Tensor Products, and Central Limit Theorems of bability Theory. VI, 107 pages. 1972. DM 16,–

273: U. Seip, Kompakt erzeugte Vektorräume und Analysis. IX, Seiten. 1972. DM 16,–

. 274: Toposes, Algebraic Geometry and Logic. Edited by. F. W. wvere. VI, 189 pages. 1972. DM 18,–

. 275: Séminaire Pierre Lelong (Analyse) Année 1970–1971. VI, pages. 1972. DM 18,–

276: A. Borel, Représentations de Groupes Localement Com- cts. V, 98 pages. 1972. DM 16,–

277: Séminaire Banach. Edité par C. Houzel. VII, 229 pages. 72. DM 20,–

Vol. 278: H. Jacquet, Automorphic Forms on GL(2). Part II. XIII, 142 pages. 1972. DM 16,–

Vol. 279: R. Bott, S. Gitler and I. M. James, Lectures on Algebraic and Differential Topology. V, 174 pages. 1972. DM 18,–

Vol. 280: Conference on the Theory of Ordinary and Partial Differential Equations. Edited by W. N. Everitt and B. D. Sleeman. XV, 367 pages. 1972. DM 26,–

Vol. 281: Coherence in Categories. Edited by S. Mac Lane. VII, 235 pages. 1972. DM 20,–

Vol. 282: W. Klingenberg und P. Flaschel, Riemannsche Hilbertmannigfaltigkeiten. Periodische Geodätische. VII, 211 Seiten. 1972. DM 20,–

Vol. 283: L. Illusie, Complexe Cotangent et Déformations II. VII, 304 pages. 1972. DM 24,–

Vol. 284: P. A. Meyer, Martingales and Stochastic Integrals I. VI, 89 pages. 1972. DM 16,–

Vol. 285: P. de la Harpe, Classical Banach-Lie Algebras and Banach-Lie Groups of Operators in Hilbert Space. III, 160 pages. 1972. DM 16,–

Vol. 286: S. Murakami, On Automorphisms of Siegel Domains. V, 95 pages. 1972. DM 16,–

Vol. 287: Hyperfunctions and Pseudo-Differential Equations. Edited by H. Komatsu. VII, 529 pages. 1973. DM 36,–

Vol. 288: Groupes de Monodromie en Géométrie Algébrique. (SGA 7 I). Dirigé par A. Grothendieck. IX, 523 pages. 1972. DM 50,–

Vol. 289: B. Fuglede, Finely Harmonic Functions. III, 188. 1972. DM 18,–

Vol. 290: D. B. Zagier, Equivariant Pontrjagin Classes and Applications to Orbit Spaces. IX, 130 pages. 1972. DM 16,–

Vol. 291: P. Orlik, Seifert Manifolds. VIII, 155 pages. 1972. DM 16,–

Vol. 292: W. D. Wallis, A. P. Street and J. S. Wallis, Combinatorics: Room Squares, Sum-Free Sets, Hadamard Matrices. V, 508 pages. 1972. DM 50,–

Vol. 293: R. A. DeVore, The Approximation of Continuous Functions by Positive Linear Operators. VIII, 289 pages. 1972. DM 24,–

Vol. 294: Stability of Stochastic Dynamical Systems. Edited by R. F. Curtain. IX, 332 pages. 1972. DM 26,–

Vol. 295: C. Dellacherie, Ensembles Analytiques, Capacités, Mesures de Hausdorff. XII, 123 pages. 1972. DM 16,–

Vol. 296: Probability and Information Theory II. Edited by M. Behara, K. Krickeberg and J. Wolfowitz. V, 223 pages. 1973. DM 20,–

Vol. 297: J. Garnett, Analytic Capacity and Measure. IV, 138 pages. 1972. DM 16,–

Vol. 298: Proceedings of the Second Conference on Compact Transformation Groups. Part 1. XIII, 453 pages. 1972. DM 32,–

Vol. 299: Proceedings of the Second Conference on Compact Transformation Groups. Part 2. XIV, 327 pages. 1972. DM 26,–

Vol. 300: P. Eymard, Moyennes Invariantes et Représentations Unitaires. II. 113 pages. 1972. DM 16,–

Vol. 301: F. Pittnauer, Vorlesungen über asymptotische Reihen. VI, 186 Seiten. 1972. DM 18,–

Vol. 302: M. Demazure, Lectures on p-Divisible Groups. V, 98 pages. 1972. DM 16,–

Vol. 303: Graph Theory and Applications. Edited by Y. Alavi, D. R. Lick and A. T. White. IX, 329 pages. 1972. DM 26,–

Vol. 304: A. K. Bousfield and D. M. Kan, Homotopy Limits, Completions and Localizations. V, 348 pages. 1972. DM 26,–

Vol. 305: Théorie des Topos et Cohomologie Etale des Schémas. Tome 3. (SGA 4). Dirigé par M. Artin, A. Grothendieck et J. L. Verdier. VI, 640 pages. 1973. DM 50,–

Vol. 306: H. Luckhardt, Extensional Gödel Functional Interpretation. VI, 161 pages. 1973. DM 18,–

Vol. 307: J. L. Bretagnolle, S. D. Chatterji et P.-A. Meyer, Ecole d'été de Probabilités: Processus Stochastiques. VI, 198 pages. 1973. DM 20,–

Vol. 308: D. Knutson, λ-Rings and the Representation Theory of the Symmetric Group. IV, 203 pages. 1973. DM 20,–

Vol. 309: D. H. Sattinger, Topics in Stability and Bifurcation Theory. VI, 190 pages. 1973. DM 18,–

Vol. 310: B. Iversen, Generic Local Structure of the Morphisms in Commutative Algebra. IV, 108 pages. 1973. DM 16,–

Vol. 311: Conference on Commutative Algebra. Edited by J. W. Brewer and E. A. Rutter. VII, 251 pages. 1973. DM 22,–

Vol. 312: Symposium on Ordinary Differential Equations. Edited by W. A. Harris, Jr. and Y. Sibuya. VIII, 204 pages. 1973. DM 22,–

Vol. 313: K. Jörgens and J. Weidmann, Spectral Properties of Hamiltonian Operators. III, 140 pages. 1973. DM 16,–

Vol. 314: M. Deuring, Lectures on the Theory of Algebraic Functions of One Variable. VI, 151 pages. 1973. DM 16,–

Vol. 315: K. Bichteler, Integration Theory (with Special Attention to Vector Measures). VI, 357 pages. 1973. DM 26,–

Vol. 316: Symposium on Non-Well-Posed Problems and Logarithmic Convexity. Edited by R. J. Knops. V, 176 pages. 1973. DM 18,–

Vol. 317: Séminaire Bourbaki – vol. 1971/72. Exposés 400–417. IV, 361 pages. 1973. DM 26,–

Vol. 318: Recent Advances in Topological Dynamics. Edited by A. Beck, VIII, 285 pages. 1973. DM 24,–

Vol. 319: Conference on Group Theory. Edited by R. W. Gatterdam and K. W. Weston. V, 188 pages. 1973. DM 18,–

Vol. 320: Modular Functions of One Variable I. Edited by W. Kuyk. V, 195 pages. 1973. DM 18,–

Vol. 321: Séminaire de Probabilités VII. Edité par P. A. Meyer. VI, 322 pages. 1973. DM 26,–

Vol. 322: Nonlinear Problems in the Physical Sciences and Biology. Edited by I. Stakgold, D. D. Joseph and D. H. Sattinger. VIII, 357 pages. 1973. DM 26,–

Vol. 323: J. L. Lions, Perturbations Singulières dans les Problèmes aux Limites et en Contrôle Optimal. XII, 645 pages. 1973. DM 42,–

Vol. 324: K. Kreith, Oscillation Theory. VI, 109 pages. 1973. DM 16,–

Vol. 325: Ch.-Ch. Chou, La Transformation de Fourier Complexe et L'Equation de Convolution. IX, 137 pages. 1973. DM 16,–

Vol. 326: A. Robert, Elliptic Curves. VIII, 264 pages. 1973. DM 22,–

Vol. 327: E. Matlis, 1-Dimensional Cohen-Macaulay Rings. XII, 157 pages. 1973. DM 18,–

Vol. 328: J. R. Büchi and D. Siefkes, The Monadic Second Order Theory of All Countable Ordinals. VI, 217 pages. 1973. DM 20,–

Vol. 329: W. Trebels, Multipliers for (C, α)-Bounded Fourier Expansions in Banach Spaces and Approximation Theory. VII, 103 pages. 1973. DM 16,–

Vol. 330: Proceedings of the Second Japan-USSR Symposium on Probability Theory. Edited by G. Maruyama and Yu. V. Prokhorov. VI, 550 pages. 1973. DM 36,–

Vol. 331: Summer School on Topological Vector Spaces. Edited by L. Waelbroeck. VI, 226 pages. 1973. DM 20,–

Vol. 332: Séminaire Pierre Lelong (Analyse) Année 1971-1972. V, 131 pages. 1973. DM 16,–

Vol. 333: Numerische, insbesondere approximationstheoretische Behandlung von Funktionalgleichungen. Herausgegeben von R. Ansorge und W. Törnig. VI, 296 Seiten. 1973. DM 24,–

Vol. 334: F. Schweiger, The Metrical Theory of Jacobi-Perron Algorithm. V, 111 pages. 1973. DM 16,–

Vol. 335: H. Huck, R. Roitzsch, U. Simon, W. Vortisch, R. Walden, B. Wegner und W. Wendland, Beweismethoden der Differentialgeometrie im Großen. IX, 159 Seiten. 1973. DM 18,–

Vol. 336: L'Analyse Harmonique dans le Domaine Complexe. Edité par E. J. Akutowicz. VIII, 169 pages. 1973. DM 18,–

Vol. 337: Cambridge Summer School in Mathematical Logic. Edited by A. R. D. Mathias and H. Rogers. IX, 660 pages. 1973. DM 42,–

Vol: 338: J. Lindenstrauss and L. Tzafriri, Classical Banach Spaces. IX, 243 pages. 1973. DM 22,–

Vol. 339: G. Kempf, F. Knudsen, D. Mumford and B. Saint-Donat, Toroidal Embeddings I. VIII, 209 pages. 1973. DM 20,–

Vol. 340: Groupes de Monodromie en Géométrie Algébrique. (SGA 7 II). Par P. Deligne et N. Katz. X, 438 pages. 1973. DM 40,–

Vol. 341: Algebraic K-Theory I, Higher K-Theories. Edited by H. Bass. XV, 335 pages. 1973. DM 26,–

Vol. 342: Algebraic K-Theory II, "Classical" Algebraic K-Theory, and Connections with Arithmetic. Edited by H. Bass. XV, 527 pages. 1973. DM 36,–

Vol. 343: Algebraic K-Theory III, Hermitian K-Theory and Ge metric Applications. Edited by H. Bass. XV, 572 pages. 197 DM 38,–

Vol. 344: A. S. Troelstra (Editor), Metamathematical Investigati of Intuitionistic Arithmetic and Analysis. XVII, 485 pages. 197 DM 34,–

Vol. 345: Proceedings of a Conference on Operator Theo Edited by P. A. Fillmore. VI, 228 pages. 1973. DM 20,–

Vol. 346: Fučík et al., Spectral Analysis of Nonlinear Operator II, 287 pages. 1973. DM 26,–

Vol. 347: J. M. Boardman and R. M. Vogt, Homotopy Invaria Algebraic Structures on Topological Spaces. X, 257 page 1973. DM 22,–

Vol. 348: A. M. Mathai and R. K. Saxena, Generalized Hype geometric Functions with Applications in Statistics and Physic Sciences. VII, 314 pages. 1973. DM 26,–

Vol. 349: Modular Functions of One Variable II. Edited by W. Kuy and P. Deligne. V, 598 pages. 1973. DM 38,–

Vol. 350: Modular Functions of One Variable III. Edited by W Kuyk and J.-P. Serre. V, 350 pages. 1973. DM 26,–

Vol. 351: H. Tachikawa, Quasi-Frobenius Rings and Generaliz tions. XI, 172 pages. 1973. DM 18,–

Vol. 352: J. D. Fay, Theta Functions on Riemann Surfaces. 137 pages. 1973. DM 16,–

Vol. 353: Proceedings of the Conference on Orders, Grou Rings and Related Topics. Organized by J. S. Hsia, M. L. Mad and T. G. Ralley. X, 224 pages. 1973. DM 20,–

Vol. 354: K. J. Devlin, Aspects of Constructibility. XII, 240 page 1973. DM 22,–

Vol. 355: M. Sion, A Theory of Semigroup Valued Measure V, 140 pages. 1973. DM 16,–

Vol. 356: W. L. J. van der Kallen, Infinitesimally Central-Exte sions of Chevalley Groups. VII, 147 pages. 1973. DM 16,–

Vol. 357: W. Borho, P. Gabriel und R. Rentschler, Primidea in Einhüllenden auflösbarer Lie-Algebren. V, 182 Seiten. 197 DM 18,–

Vol. 358: F. L. Williams, Tensor Products of Principal Seri Representations. VI, 132 pages. 1973. DM 16,–

Vol. 359: U. Stammbach, Homology in Group Theory. VIII, 18 pages. 1973. DM 18,–

Vol. 360: W. J. Padgett and R. L. Taylor, Laws of Large Numbe for Normed Linear Spaces and Certain Fréchet Spaces. 111 pages. 1973. DM 16,–

Vol. 361: J. W. Schutz, Foundations of Special Relativity: Kin matic Axioms for Minkowski Space Time. XX, 314 pages. 197 DM 26,–

Vol. 362: Proceedings of the Conference on Numerical Solutic of Ordinary Differential Equations. Edited by D. Bettis. VI 490 pages. 1974. DM 34,–

Vol. 363: Conference on the Numerical Solution of Differenti Equations. Edited by G. A. Watson. IX, 221 pages. 1974. DM 20

Vol. 364: Proceedings on Infinite Dimensional Holomorph Edited by T. L. Hayden and T. J. Suffridge. VII, 212 pages. 197 DM 20,–

Vol. 365: R. P. Gilbert, Constructive Methods for Elliptic Equ tions. VII, 397 pages. 1974. DM 26,–

Vol. 366: R. Steinberg, Conjugacy Classes in Algebraic Grour (Notes by V. V. Deodhar). VI, 159 pages. 1974. DM 18,–

Vol. 367: K. Langmann und W. Lütkebohmert, Cousinverteilu gen und Fortsetzungssätze. VI, 151 Seiten. 1974. DM 16,–

Vol. 368: R. J. Milgram, Unstable Homotopy from the Stab Point of View. V, 109 pages. 1974. DM 16,–

Vol. 369: Victoria Symposium on Nonstandard Analysis. Edite by A. Hurd and P. Loeb. XVIII, 339 pages. 1974. DM 26,–

Vol. 370: B. Mazur and W. Messing, Universal Extensions ar One Dimensional Crystalline Cohomology. VII, 134 pages. 197 DM 16,–